Digital Radio in Europe

Digital Radio in Europe
Technologies, Industries and Cultures

Edited by

Brian O'Neill, Marko Ala-Fossi, Per Jauert, Stephen Lax,
Lars Nyre & Helen Shaw

intellect Bristol, UK / Chicago, USA

First published in the UK in 2010 by
Intellect, The Mill, Parnall Road, Fishponds, Bristol, BS16 3JG, UK

First published in the USA in 2010 by
Intellect, The University of Chicago Press, 1427 E. 60th Street,
Chicago, IL 60637, USA

Cover designer: Holly Rose
Copy-editor: Michael Eckhardt
Typesetting: Mac Style, Beverley, E. Yorkshire

ISBN 978-1-84150-279-3

Printed and bound by Gutenberg Press, Malta.

Contents

Acknowledgements

The editors would like to acknowledge the support of the Dublin Institute of Technology and the University of Bergen in the publication of this volume. Thanks also to Michael Mullane at the European Broadcasting Union for the invitation to present at the EBU's Digital Radio Conference in 2007 and 2008, and for helpful dialogue on the research. And especial thanks to May Yao and James Campbell at Intellect Books for support and guiding this project to publication.

Foreword

The Ontology of Radio

Paddy Scannell

The ontology of broadcast radio (what it *is*) concerns its *listenability*: a technical issue, a pragmatic problem, an ontological phenomenon.

Technical matters

Technologies are, in the first instance, technical but in the last instance they are not. Radio, as a technology, was a scientific discovery of the later nineteenth century whose applications were taken up, explored and exploited by an emergent electronics industry. Broadcast radio was discovered, after World War 1, as a byproduct of a technology conceived of as an extension of wired telephony, and first used for military, maritime and business purposes. What started as a device for two-way talk between distant parties unconnected by wire became a one-way system of transmission in which broadcasters provided something for the purchasers of radio sets to listen to.

As this collection makes clear, the switch from analogue to digital radio broadcasting in Europe involves scientific innovation, industrial uptake and political regulation at national and trans-national levels. All this comes before the application and uptake of the technology in the production, transmission and reception of radio programmes, and yet that is their driver. For what determines the relations of production in any communications technology is always ultimately its fundamental utility or use. Books are to be read and hence all aspects of their manufacture are determined by consideration of their readability (their ease and convenience of use). Television production is determined by considerations of its watchability, and radio by its listenability. Let us consider the question of listenablity first as a technical problem for that is how it first appeared. Radio listening in its early years was a hit and miss affair: the signal strength fluctuated and its range was limited. Transmissions from other stations on the same or similar wavelength interfered with what listeners had tuned into, and oscillation from other nearby radio sets was a menace in the 1920s. Technical innovation in audio technologies in the radio and music industries has been, from the start, concerned to improve the quality of the listening experience for the purchasers of gramophones, wireless and their contemporary equivalents. The elimination of interference, the achievement of clear, strong signals and the auditory quality (clarity, tone, timbre etc) of the sound transmitted by the receiving apparatus have been constants from the beginning of the audio industries to now. Digital audio broadcasting (DAB) is a new solution to a very old problem that never goes away – maintaining the quality of the listening experience.

From a technical point of view that means maintaining the purity (the high fidelity) of the audible sounds (the acoustics) as broadcast so that there is no loss of quality in transmission. What is transmitted from the recording or broadcasting studio and what comes out of the receiving apparatus has suffered no loss or distortion. It may even have been enhanced by high fidelity and stereo equipment. All this has a long, continuing history that is utterly invisible both in the experience of radio for listeners and (until very recently) its academic study. What are we to make of this, the invisibility of the technology itself in its application and use? It is best thought of, I suggest, as its gift. What technologies give their users is that they may use the thing (the book, the TV or radio set, the iPod player) without being aware of the thing they are using. We only become aware of technologies when they go wrong; when, in the case of radio, the signal starts to break up or fade, or to sound mushy or tinny. Technologies of communication are designed to foreground the communicative experience by eliminating, as far as technically possible, all those things that might come between the listener, viewer or reader and what they are listening to, watching or reading. In so doing, they render themselves invisible. We do not notice them. What they grant is that we can take them for granted. This is the sought for and intended outcome of all the labour that is hidden in technologies; all the human research, experimentation, trial and error that is their *care structure*. And so it is with the production care structure of radio.

Pragmatic considerations

It too has a care to conceal its own practices which are all oriented to the listenability of what is broadcast. Once technical issues have been brought under control (a paramount concern in the very early history of radio), so that the acoustics of listening are relatively unproblematic, what then comes to the fore is the central communicative problem of production: how to broadcast something that listeners might, in fact, want to listen to. All radio output is reducible to two basic categories: music and talk. The microphone is radio's basic communicative technology, and its acoustic properties have profound consequences for the ways in which the human voice is deployed as it sings or speaks in front of it. Music is the easiest way of filling time on radio and from the start has always constituted the bulk of what is transmitted, everywhere. When singers came into radio studios in the early years of broadcasting to perform at the microphone it was immediately found that they had to stand about six feet away from it otherwise they would shatter windows and the eardrums of listeners. Singing in public at the time required voices capable of filling a large auditorium, whether an opera house or a music hall, without the use of any technology to amplify the sound produced by the singer. Hence all performers in public learned to pitch their voices loud and strong for a large audience in a large public space and, in the case of opera, over the sound of a large orchestra. The microphone transformed the proxemics (the social space and hence the social relations) of singing.[1] In the thirties singers at the microphone discovered that they could stand close to the microphone and lower their voice instead of standing back and bawling at it. A new technique of 'close-mike' singing was developed in the USA and Britain, labeled crooning at the time. This was not exactly a term of endearment and the new style was controversial on both sides of the Atlantic, but it did change how singers could use their voices and the experience of listening to singing.

The microphone transforms the social dynamics between the singer, the song and the listener, bringing all three closer to each other and thus into a more intimate, personal relationship in which the singing voice speaks to an individual person rather than bellows at a large collective gathering in a large public space. Singing is personalized. The singer now performs in his or her own 'natural' voice, instead of the pervasively 'unnatural', artificial operatic voice that characterized the dominant style of singing in public before the coming of the radio. In the 1940s you heard the voice of Frank Sinatra as the voice *of* Frank Sinatra; that is you heard him singing in his own voice (not an artificial, stage voice) and as the person that he was, and you heard his song as if it spoke to you, personally. Close-mike singing is conversational in style (it implicates a 'you-and-me' relationship). The words (what the song is saying) become more important. The voice, as it is lowered, as it comes closer to speaking or even murmuring, is heard as less artificial and more natural. It also is more expressive. It lacks the remarkable aesthetic range of operatic voices – nearly all close-mike singers are baritone or contralto – but what it lacks in vocal range it makes up for in a quite new range of expressiveness. More pathos, humour, sensuality and sexiness – in short, more personality and mood – can be projected by the singer crooning into the mike than the stage performer projecting out into the auditorium.[2]

What is true of the social relations of public singing is equally true of the social relations of public speaking. What was 'talk in public' before radio and television? It largely consisted of a speaker and an audience in a defined public space. The lecture, the sermon and the political rally – these were the most familiar forms of public discourse in the early decades of the twentieth century. In all of them a public speaker addressed a crowd. Each had a distinctive performative rhetoric, a style of speaking, that marked it as the lecture, the sermon the political harangue, and underlined the status of the lecturer, the preacher, the politician as distinct from his (and public speakers were invariably male) massed audience of indistinguishable individuals. Again, as in styles of public singing before radio, public speakers must make themselves heard at the back of the assembly they addressed and they spoke as public persons, not as themselves, to an undifferentiated massed audience and not a constellation of individuals.

In Britain it took the BBC Talks Department some time to learn how to talk to its absent audience of listeners. None of the existing models of public speaking were found to work and this was for one overwhelming reason: the circumstance of listening to the radio did not correspond with the circumstances of listening to a lecture, a sermon or a political harangue. People listened to radio in their homes as part of domestic daily life. In that situation and those circumstances listeners did not want to be lectured, preached at, harangued or generally 'got at'.[3] They wanted to be addressed by the radio in ways that were appropriate to the situations and circumstances in which they were listening. Broadcasters soon learnt to adopt a conversational style of address, to talk to the listeners 'out there' as if they were an audience of one, to draw them into a conversation.

Conversation is a distinctive kind of talk. Perhaps its defining characteristic is best understood negatively. It is not institutional talk which pre-defines the allocation of speaker roles and the distribution of who says what. In the classroom, the surgery or the law court one class of speakers controls and defines the interaction: teachers, doctors and lawyers ask questions and pupils, patients and plaintiffs answer them. Conversation in non-institutional situations and settings is a jointly managed interaction in which speaker-listener roles are evenly distributed and participants are equally involved in the business of starting it, keeping it going and closing it down. It presupposes a shared responsibility for the management of the interaction. Something of this is implicated in the new style of talk developed on radio and television, and made possible by the communicative affordances of the microphone (what it allows for and facilitates).[4] To be sure, all talk on radio and television is institutional. Responsibility for what is said, first and last, resides always with the broadcasters who control the management of talk on radio and television. But it is oriented towards the norms of ordinary conversation in non-institutional settings, in everyday situations and circumstances because that is where it is heard and responded to.

Thus the pragmatics of music and talk on radio were from the start and continue to be defined by the microphone which, like every technology, has a long hidden history of research, innovation and refinement to produce smaller, more sensitive and portable instruments for different purposes in different locations. But always its overall effect, its communicative affordance, is to make possible more natural and less artificial performative styles in singing and speaking. Thus radio's characteristic mode of address is not one in which speakers shout as if to an audience at the back of a large hall or lecture theatre, nor one in which they whisper as if in your ear. Broadcasters *talk* as if the listener is within the social space of a conversation at which he or she was a present participant. Shouting implicates an impersonal distant relationship, whispering a personal intimate close-up relationship, and talk an intermediate inter-personal relationship. The social proxemics of radio, a direct effect of the technical affordances of microphones, are oriented to communicative ease or sociability. This seems to be a necessary consequence of the unforced communicative relationship between broadcasters and listeners.

In relations of presence, when we engage in interactions they are binding for their duration. This is evidently so in public, institutional situations: whether it's a concert, a play, a lecture or a sermon audiences are bound to observe what Goffman (1972) calls the 'situational proprieties of the occasion', and if they do not they will be sanctioned (usually by expulsion). And even in ordinary talk in everyday circumstances (including, or perhaps especially, on the phone) we are required to show interest, to be attentive, actively to engage in the work of keeping the conversation going without lapsing into awkward silences or embarrassment. None of this applies in the social relations between broadcasters and their absent audiences. It is a cardinal consideration for broadcasters to create and maintain at all times communicative ease (a sociable mood) between themselves and their listeners. In relations of presence, audiences must adjust their dispositions and demeanors to what the performance and the performers require of them. In the relations of radio broadcasting this is reversed, and performers and performance are always adjusted to the circumstances of the audience. The communicative relationship that the microphone facilitates is non-dominative because it cannot be enforced (you can't *make* anyone listen to the radio) and presupposes equality of being between broadcasters and listeners. It is very easy to hear if you are being lectured or condescended to or somehow manipulated, and listeners have an instant remedy. Their power lies in the off-button– as broadcasters know and understand full well.

The phenomenology of hearing

Having attended in the first place to the necessary technical matters that give the possibility of listening at all, and then to the pragmatic issues at stake in producing something that is listenable to, we are now in a position to raise the question: what is it that we *hear* in listening to the radio? More exactly, what is it that *gives* what we get to hear? What else but the human voice as it speaks and sings. Voice is the universal medium of speech; our human ability or capacity for language. Language has three distinctive functions or uses: it is informative, communicative and expressive. The informative function of language is about statements of fact; truth as the correspondence between well formed sentences and worldly states of affairs ("Snow is white", for instance) – the object domain of much twentieth century analytic philosophy. The communicative function of language explores its social, relational and interactive character. And then there is the expressive dimension – the most elusive, least analyzed and analyzable of the three functions of language. The power of speech is not an effect of language but is anterior to it. It resides not in what is said, nor any content of spoken utterances or written sentences. It is primarily an effect of the human voice as the expressive register of our being. It is so in a number of ways that have a heard but unnoticed character. In any human voice others will hear aspects of who the speaker *is*: their sex, age, education, which country they come from (Ireland or America, say) and, if you have a well tuned ear for regional accents, which part the country they come from. These are general aspects of any individual identity, but voice also identifies you in particular; you as the person that you are. We recognize others as individuals, inter alia, through the individual sound of their voice, as work on telephone talk has shown. Radio listeners quickly learn to recognize, because they must, who is speaking at the microphone through the sound of their voice alone. A long running drama serial like *The Archers* (BBC, Radio 4) depends entirely on voice recognition for its intelligibility from moment to moment in every episode.

Voice is disclosive of all this and more, for it is not only the register of our social being (who we are), it is also the expressive register of our existential being; our *being* sad or happy, interested or bored, tired, excited, edgy or angry… all our ever-shifting moods (our states of being) as disclosed in and by the endlessly nuanced, infinitely suggestive, revelatory play of the human voice as it speaks and sings. What we hear whenever we listen to the radio is the sound of life, disclosed in the *liveness* of voice. This is irrespective of whether the transmission is, technically, live-to-air or recorded. We never hear the voices on air as disembodied, ghostly phenomena. We always hear them as embodied utterances of real live people (whether factual or fictional) in the living, unfolding moment of the broadcast 'now'. We hear this not as if it were elsewhere ('there') but as if 'here' where we are as we listen. If radio seems to us to be part of our daily lives it is because that is how we hear it. The embodied liveness of voice expresses the essence (the being) of *live*, if by that one means our being in the here and now, in the living moment of the enunciatory speech act. Ontology, the *logos* of *ontos*, the discourse of being, is something that we, in the first place, hear and attend to in the play of the human voice as the expressive register of a communicable sharable experience of being in the world. Something of this is what we hear in unmarked ways whenever we listen to the radio.

Notes

1. Proxemics is a term coined by the anthropologist, Edward T. Hall, to capture the social organization of space in all its cultural variety in everyday situations. See Hall 1966.
2. On these developments in the UK see Scannell 1996: 58–74. For the USA, see McCracken forthcoming.
3. This is how Hilda Matheson, the first Head of Talks at the BBC (1927–1932), discusses the matter in her seminal insider account of the impact of radio (Matheson 1933: 75 ff.) For a more detailed historical account of the management of talk on radio, see Scannell and Cardiff 1991: 153–180.
4. On the communicative affordances of technologies, especially in relationship to talk, see Hutchby 2001.

References

Goffman, E. (1972), *Interaction ritual*, Harmondsworth: Penguin Books.
Hall, E. (1966), *The hidden dimension*, New York: Anchor Books.
Hutchby, I. (2001), *Conversation and technology*, Cambridge: Polity Press.
McCracken, A. (forthcoming), *Real men don't sing. Crooners and American culture, 1925–33.*
Matheson, H. (1933), *Broadcasting*, London: Home University Library.
Scannell, P. (1996), *Radio, television and modern life*, Oxford: Blackwell.
Scannell, P. and D. Cardiff (1991), *A social history of British broadcasting, 1922–1939*, Oxford: Blackwell.

Introduction

At a fundamental level, radio is a medium that has changed little over the course of a century. The system of wireless transmission, using continuous radio waves to carry analogue signals of voice and music, to be received by potentially millions of listeners via electronic sets, is pretty much the same system that pioneers like Guglielmo Marconi and Reginald Fessenden created and used in the early years of the twentieth century. Radio over its history has apparently adapted to change in an effortless way. Initially through transistorisation and then through better transmission methods and improved receiver design, radio became portable, mobile, an indispensable feature of in-car entertainment, a personal companion and a feature of almost every room in the typical home environment. Challenges to radio – whether from television, new recorded music formats, changing patterns of media consumption – have all threatened at some time to supplant radio as a communications medium. Yet radio's ability to survive, to adapt to change and to remain relevant on a global scale is undisputed and remarkable.

Radio is, however, now undergoing renewed challenges and significant change. Digital technologies and the growing convergence of technology and media platforms through mobile communications and the Internet have radically transformed the environment in which radio, as traditionally conceived, operates. The next generation of radio listeners has grown up with very different media habits and despite the fact that radio in Europe appears to be a thriving and vibrant cultural force, with tens of thousands of stations and over two hundred million listeners who spend about three hours every day listening, its long term future and viability cannot be taken for granted. This was the view expressed by Jenny Abramsky, BBC's former Director of Radio & Music, speaking to the National Association of Broadcasters in 2003 when she argued 'radio must go digital if it is not to go into long term decline. If radio were the only medium not to go digital it would soon become obsolete for younger audiences and future generations accustomed to a vast array of digital choices when it comes to their media experiences'(Abramsky 2003).

The aim of this edited volume, *Digital Radio in Europe: Technologies, Industries and Cultures*, is to critically examine this process of change and re-invention. Bringing radio into the digital era means more than simply changing its mode of delivery into digital form. As audiences are aware from television's transformation in the digital era, major changes in the medium have taken place in how television is produced, how it is displayed and presented, and ultimately how we as audiences use the medium. Radio is, of course, a very different medium. The same forces that, for example, have led the European Commission to recommend that member-states switch-off analogue television by 2012, do not apply in radio. Its transition to a fully digital form has been much more gradual and incremental. The research collected in this volume examines the histories, competing technologies, and opportunities and challenges that face broadcasters as they seek to support radio's drive to become digital. Drawing on a variety of perspectives from media studies, technology studies, media policy and industry analysis, we explore the original political vision which gave rise to digital radio in the European context. We analyse the different strategies and policies that have been pursued in individual countries and regions, comparing contexts in which digital radio has had a measure of relative success against those in which digital radio has demonstrably failed. We also explore how changing and evolving forms of radio may impact on listeners and audiences. Will new forms of digital radio offer new opportunities for audience access and participation, and for an enhanced public sphere, for instance? Can the potential of new sound media be harnessed for greater social benefit? Or is the challenge of digitalization ultimately a threat to the special qualities of intimacy and localness that radio as a medium has developed over its history? The testing of radio's resilience, and its ability to evolve into the digital age as a result, raise important and fundamental issues of public value that we believe have importance and relevance not only for specialists in radio, but for practitioners, policy-makers and audiences of all media.

From analogue replacement to multimedia convergence

The original impetus for the development of a digital radio system in the early 1980s was to find a replacement for analogue radio transmission, that is, the use of either amplitude (AM) or frequency modulation (FM) in place since the early part of the twentieth century. Radio, it was claimed at the time, needed to take a step forward to keep pace with technology developments in other field of communications and to overcome problems of interference, difficulties of tuning and limited capacity on analogue systems. A digital broadcasting system, on the other hand, it was argued, would bring major enhancements such as greatly improved audio quality, intuitive and precise tuning, an increase in channel capacity and a range of new data and interactive services. After all, most other aspects of the audiovisual production chain are in digital form, including the various different forms of digital audio recording, production, storage, digital signal processing techniques and the host of digital media entertainment formats and platforms that make up the lucrative consumer electronics sector.

Scientific research into developing a replacement technology for radio in Europe, particularly for FM, began in earnest in the mid-1980s with the formation of a consortium of leading European broadcasters and research institutes, supported by the European

Broadcasting Union. This project, supported under the Eureka European research and development funding programme, became known as Eureka 147 and developed the Digital Audio Broadcasting standard, or DAB as it is widely known.

DAB is now the longest established and most mature technology for digital radio. The first successful stations using DAB digital technology were launched in 1995. Internationally, there are now over 1000 services available in over 30 countries, and more than 12 million DAB receivers have been sold worldwide (WorldDMB 2009). Regular DAB services have been launched in a number of European countries with the ultimate goal of moving analogue services onto the digital platform. Notable DAB successes have included the UK with a receiver base of 8 million digital radio sets, and over 10% share of all radio listening (RAJAR Q2 2008). In Demark similarly, strong support from the government and the public broadcaster has resulted in an extensive roll-out of digital radio. Other European countries offering regular DAB services include Belgium, Germany, Ireland, Netherlands, Norway, Portugal and Sweden. Countries which offer more limited services or which are piloting newer technologies such as DAB+ or DMB include France, Italy and Austria.

The original ambition of the Eureka 147 consortium was that DAB would be a world standard, and would put European technology companies in a position of global leadership for the roll out of new consumer electronics systems and products. This ambition, as we discuss in the following, has not been realised however, and DAB is not the only digital radio standard or the only means of bringing radio to the public in digital form. Competing standards for digital radio have been developed in the United States and Japan, for instance, and very different strategies have been employed for migration of services to a digital broadcasting environment. Some twenty years after its original development, DAB is just one of a number of different digital platforms offering audio and multimedia services operating in a dynamic and fast-changing media market.

In addition, the pace of progress of radio's move into a digital environment has been slow, and a complete migration to a digital system and switch-off of analogue transmission networks remains a distant prospect. By contrast, systems to replace analogue television transmission are well established and the process of migration is solidly underway. Digital television transition is a priority for governments everywhere driven by the fact that converting to digital will yield a rich dividend by freeing up large amounts of valuable spectrum. Deadlines have been firmly established and the process of analogue television switch-off has already occurred or is in process, with most Western European countries planning to terminate transmission on or around 2012. The transition to digital television services has been rapid, requiring consumers to install new receiving equipment while availing of an ever-increasing array of information and entertainment channels options across DTT, digital cable, satellite and IPTV platforms. Intense competition has led to the provision of a host of new services, including interactive features and the bundling of alternative communications services, such as data and voice, as part of the television package.

The same is not the case for digital radio and despite the many years of investment in research and technology development, and the high priority given to a digital transition by some broadcasters and government regulators, there is probably less consensus now about the future of radio than at any time in the past. The once straightforward proposition of updating the transmission system, comparable to the transition from AM to FM at an earlier stage in radio history, is now a much more complex problem, complete with competing options and platforms, fragmentation in the market place and uncertainty among radio broadcasters and regulators. Road maps for a digital transition have been established in major markets such as France and the UK. In France, a timetable has been established which will require all new radio receivers – including in-car and mobile phones receiver – to be digital-enabled by 2013. In the UK, the government has set 2015 as the target date by which Britain's national radio stations, and most local stations, will stop broadcasting on analogue and move to Digital Audio Broadcasting. The UK's *Digital Britain* report commits the government to a policy 'enabling DAB to be a primary distribution network for radio' and creates a plan for a digital switchover once there is sufficient national coverage and 50% of radio listening is digital. Despite this, uncertainty remains. There are frquent disparaging newspaper headlines suggesting that DAB may turn out to be the 'Betamax' of radio (Plunkett 2008) and predictions that the enduring appeal of analogue radio, combined with the growing power of Internet radio, may ultimately lead many consumers to sidestep DAB technology altogether (Deloitte 2009).

Online and other digital networks

Digital radio, as we argue in this volume, has now come to mean more than just the replacement of analogue broadcasting with a digital equivalent. In practical terms, radio now exists as a multi-platform medium with analogue, digital, satellite and the Internet being the most important of its delivery platforms. With the rise of the Internet, the greater availability of broadband, and the rapid expansion of mobile communications technologies, there are now numerous alternative ways of providing radio services to listeners in digital form. Live audio streaming via the World Wide Web and podcasting have become highly successful alternative means of distributing radio on the Internet. Internet-only radio stations and the development of personalised Internet radio services such as *Rhapsody* or *Last.fm* have also provided significant challenges to traditional broadcasting. Rapid advances in mobile communications and the development of multimedia services for mobile devices likewise compete with radio as outlets for digital audio services, making the market for digital radio a highly complex one.

There is, as a result, strong commercial pressure on the radio industry to fully engage with digital convergence in the media marketplace. Radio, as noted in the 2006 European Commission study of the digital content industry, is often overlooked when thinking about convergence and interactive media (Screen Digest 2006). Online music distribution, by contrast, has developed into a major new industry expected to be worth €1.1bn by 2010 and three times that in the United States. The same study estimated that there are currently

15 million listeners to online radio in Europe, expected to reach 32 million listeners or 7% of Europeans by 2010, and a further 11 million listeners for podcasting also by 2010. Against this, the total revenue anticipated by 2010 for all digital radio will be just 5% of the overall advertising revenues for the radio sector, and as a result there is major pressure on the industry to find ways to ensure it builds a higher profile in the digital content market.

Digital radio as a concept now stands at a crossroads in its development, leading some to question whether it has a viable future or whether radio as such is more likely to converge with other forms of multimedia broadcasting and personalised audio media services. As it stands, radio is poised on the cusp of a rapidly expanding environment for digital media services where it can potentially contribute to diverse platforms, including handheld mobile devices, online streaming and download services, and multimedia-rich cable and satellite services. This represents a considerable metamorphosis of radio as traditionally conceived in the one-to-many broadcasting model to stand-alone receivers. It is also quite different to the original conception of broadcast digital radio as conceived by the developers of the DAB Eureka 147 standard. Despite the mixed level of success for digital radio, industry advisors maintain a consensus that radio must go digital and interactive if it is to survive in the twenty-first century. Radio, it is argued, needs to embrace interactivity, digitalization and the Internet just as television and other media have done, and incorporate such features as search engines, digital recording and mobile broadcasting in order to stay current and relevant to today's audiences.

Technologies, industries and cultures

As Paddy Scannell reminds us in 'The Ontology of Radio', the foreword to this collection, radio technologies are only in part a technical matter and their fundamental utility or use ultimately shapes radio as a medium. The essays collected in this volume, therefore, are organized around three main areas of discussion, examining in turn issues relating to the *technologies* by which digital radio is delivered and received; the *industry contexts* which define organizationally the strategies, policies and professional practices of current digital radio; and the *cultures* of listening and participation, including those that have been observed within contemporary digital radio culture and those that may evolve from emerging forms of cultural practice.

Technologies of digital radio

With regard to the technologies of digital radio, we offer a critical review of the different platforms for digital radio delivery and reception, their origins and social implications. In Chapter One, O'Neill and Shaw examine the original aims and evolution of the Eureka 147 standard within the context of a European public service vision for a common digital future, and analyze some of the factors that have shaped the development of the DAB technological system. Ala-Fossi extends this analysis of digital platforms in Chapter Two by providing a comprehensive survey of the technological landscape of radio in the twenty-first century. Assessing the competing technological options for digital audio delivery in a global context, Ala-Fossi questions the extent to which digitalization has been a force for convergence in the radio media industry and argues that there are now a number of competing scenarios for radio in the digital era. In Chapter Three, Lax returns to the origins of the Eureka 147 DAB standard and assesses the promised innovations of digital radio in a comparative historical context, citing the introduction of stereo FM as a more radical form of technological change. O'Neill follows this in Chapter Four with a critical examination of digital radio's claim to be 'the sound of the future', despite the fact that audio quality has often been sacrificed in favour of a greater quantity of services.

Industries

Chapters Five to Eight locate digital radio in the context of Europe's large radio industry, and provide detailed analyses of the regulatory and policy contexts in which digital radio has evolved, as well as the distinct strategies towards digitalization that have been adopted in a variety of regions and markets. The core of this analysis is in Chapter 5, where Jauert et al compare the policies that different European countries have introduced to support the introduction of DAB, and examine the individual roles played by governments, regulators and public service broadcasters. Contrasting case studies illustrating relatively successful implementations such as the UK and Denmark, alongside marked failures such as in Finland, highlight the very uneven nature of digital radio progress in Europe. Despite setbacks, evident in the slow pace of progress in rolling out the first generation technology of DAB, renewed efforts to launch digital radio on the updated DAB+ standard are underway, and these are discussed within the context of the policy incentives and consumer interest required for success. Chapters Six and Seven consider digital radio technology policy from a comparative perspective and examine case studies in the United States, in Canada and Australia. Stavitsky and Huntsberger analyze in Chapter Six how in the United States, despite early interest in DAB, an alternative, in-band system, HD Radio, was developed as the approved digital radio standard. O'Neill, in Chapter Seven, compares Canada, an early supporter and adopter of the European DAB system with Australia, as the first country to formally launch DAB+ on a national level as the digital radio standard for its radio industry. In Chapter Eight, Ala-Fossi presents a survey of industry expert opinion on the future of radio, on why certain technologies are seen as more promising, and which delivery technologies radio professionals consider will be most successful. Four different scenarios for the future of radio media are considered and mapped against industry priorities and strategies.

Cultures

The cultures of digital radio consumption and participation provide the third thematic focus for this collection within which authors examine the cultural context for new modes of reception and audience engagement with the medium of digital radio. The objective of the enquiry is to analyze the distinct cultures of consumption that have arisen in the context of digital radio, and explore ways in which digital radio can enhance listener experiences, improve audience access and participation, and contribute to a more democratic media environment. In Chapter Nine, Hallett examines what digital radio has to offer the community radio sector, and argues that it is only with second generation digital broadcast platforms that there are any benefits apparent for this very important tier of radio. In Chapter Ten, Nyre and Ala-Fossi assess the thesis that digital radio extends the traditional broadcasting paradigm and enables greatly increased interaction and direct involvement of listeners. They argue, however, that the use of digital technologies for greater participative public communication poses the dilemma of the loss of privacy and increased audience surveillance. Whether this constitutes a major barrier to participation is considered through comparative case studies of audience interaction within the digital public sphere across a number of European counties. Finally, Shaw in Chapter Eleven examines radio's online transformation and how different forms of Internet radio are promoting new practices of consumption, challenging assumptions of what radio is and what it might become in the future.

Extending radio – improving radio?

Given the twenty five years of innovation and development in European digital radio that has taken place, an overall concern in this volume is the degree to which the basic format of radio has been extended and whether this has enhanced radio's potential as a medium of communication. Digital developments have brought many enhancements to the underlying technology of radio. The DAB format, for example, introduced graphic and visual elements which open up new avenues of communication; it also introduced simple menu-driven tuning, enhanced sound quality, and the promise of interactive features. Internet radio has also been a site of intense experimentation from the mid-1990s on, with web radio, file-sharing and podcasting, combined with varying levels of visual content, stretching the boundaries of what we understand as radio.

Although the technical functionality of radio may have been extended, the extent to which it has substantially changed or improved radio in *practical* or *editorial* terms remains a matter of debate. Traditional forms of broadcasting continue to dominate on new digital platforms, with little obvious exploration or experimentation with the communicative novelties that new media might offer. It is illustrative, for example, that the enthusiasm for media experimentation on social networking sites like YouTube and MySpace is not well catered for within established European radio formats, or that the radical opportunities for public participation that the Internet offers aren't further incorporated within the radio production process.

Background to the research

This collection has its origins in research developed by the *Digital Radio Cultures in Europe* research group (DRACE).[1] DRACE is an international research group comprising communication scholars and radio professionals in Denmark, Finland, Norway, Ireland and the United Kingdom, with a particular research interest in the transition of radio from analogue to digital form. DRACE began as a working group within a larger European research initiative on 'The Impact of the Internet on Mass Media', a thematic network funded under the COST programme (*Cooperation in the field of Scientific and Technical Research*). The purpose of COST Action A-20 was to develop knowledge about the various changes that mass media industries – radio, television, newspapers and cross-media – are undergoing, and will undergo in the future.[2] Our original research built on initial studies of the use of websites in radio and the evolution of DAB as a digital platform, to a wider consideration of the emerging technological landscapes of radio and overall patterns of access and participation in the digital environment. The research conducted by DRACE has included extensive cross-national technology foresight studies, contemporary policy and market analysis, and historical studies of the emergence of digital and online radio from the 1980s on. Extensive interviews were carried out with industry professionals in Europe and North America about journalistic/editorial strategy, business models and commercial and public service in the digital environment. DRACE researchers have also conducted qualitative audience research, focus groups and surveys in several European countries exploring the reception of digital radio platforms and the cultures and contexts in which they are used. Results across the different individual research projects have largely been set in a comparative perspective through examination of the contexts for digital radio in countries such as the UK, Canada, Denmark, Finland, Ireland and Norway.

As authors and editors, our purpose in presenting the research in this collection is threefold. Firstly, our objective is to contribute well-grounded empirical studies into the debate about radio and new technologies. To date, much of the discussion about digital radio has been industry-led and geared to the needs of stakeholders. Detailed debate about digital radio has been restricted to technical or scientific groups, specialist trade press and to professional radio industry groups. Within this discussion, there has been a tendency for hype about digital technologies in general and, to a certain extent, manufacturer-originated marketing about the merits of individual platforms. There has been correspondingly little policy-oriented or media studies attention given to the issues involved. Little or no research from this perspective has been undertaken with respect to audiences' needs, practices and expectations. This is a major gap which we believe requires urgent attention.

Secondly, it is our contention that the debate about the future of radio in the digital environment raises issues of central public importance, and as such we seek to contribute to a greater *public understanding* of what is involved. Our focus, as articulated above, deals with what is gained and what is lost in the transition from analogue to digital. As the oldest broadcast communications medium, radio has played and continues to serve a crucial function in constituting a public sphere of both local and global dimensions, informing and unifying citizens in communities around the world. Public policy on the future of the medium impacts, therefore, in a very direct way on people's everyday lives. Insufficient attention has been given, we believe, to this public dimension of the debate about digital radio. An underlying assumption of our research is that communications media constitute a public good and that the public interest in the emerging digital media environment needs to be closely monitored and defended. Opportunities for public participation in this debate have, to date, been wholly inadequate.

A third objective behind the research in this project is to point towards newer and better strategies and approaches in digital radio media. Rather than just serve further consumer demand with more and more services, or extract further value from new technologies in the audiovisual marketplace, there are, we believe, ways in which new audio media platforms can be deployed that enhance the role that media play in society, and contribute to a better, and more democratic public environment.

Digital Radio in Europe: Technologies, Industries and Cultures contributes to a growing body of literature on European media and communications policy, dealing with the competing and sometimes contradictory trends, interests and information about the impact of technologies on the wider European public. Radio to date has been noticeably absent from such discussion and it is hoped that that the present volume will highlight the distinctive and important issues concerned.

Notes

1. See the DRACE website at: www.drace.org .
2. See website of COST A20 at: http://cost-a20.iscte.pt/

References

Abramsky, J. (2003), 'The future of digital radio in Europe', *Speech to the National Association of Broadcasters*, 20 October 2003, http://www.bbc.co.uk/pressoffice/speeches/stories/abramsky_nab. shtml. Accessed 25 May 2009.

Deloitte (2009), 'Digital Audio Broadcasting (DAB) radios – the Digital Index', http://www.deloitte. com/view/en_GB/uk/industries/tmt/making-the-digital-transition/the-digital-index/dab-radio/ index.htm. Accessed 21 July 2008.

DCMS (2009), *Digital Britain: The Interim Report*, Department for Culture, Media and Sport, http://www.culture.gov.uk/what_we_do/broadcasting/5944.aspx. Accessed 21 July 2008.

Plunkett, J. (2008), 'Is DAB radio the next Betamax?', *The Guardian*, Tuesday 29 January 2008.

Radio World (2009), 'French Radio Sets All Digital by 2013', http://www.radioworld.com/printableView.aspx?contentid=76464. Accessed 21 July 2008.

Screen Digest Ltd, CMS Hasche Sigle, Goldmedia Gmbh, Rightscom Ltd (2006), *Interactive content and convergence: Implications for the information society*, European Commission, Information Society and Media.

WorldDMB (2009), 'Introduction to Digital Broadcasting', http://www.worlddab.org/introduction_to_digital_broadcasting. Accessed 21 July 2008.

Chapter 1

Radio Broadcasting in Europe: The Search for a Common Digital Future

Brian O'Neill & Helen Shaw

B roadcasting services, as the media historian Burton Paulu observed in the mid-1960s, are integral parts of the countries they serve (Paulu 1967: 5). European broadcasting consequently mirrors great geographical, political, linguistic, cultural and religious contrasts, and constitutes a polyglot patchwork of competing cultural interests as diverse as the continent itself. Yet there is also an underlying tradition and an historical experience in common, particularly evident in the case of radio. The European consensus that broadcasting should be protected from purely commercial pressures, and that it should be subject to unified control and organized as a monopoly, was established early on (McDonnell 1991: 1), and has left a legacy of state-chartered corporations, predominantly funded by licence fees, with a tradition of strong, mutual co-operation and commitment to high programme standards. It is also a tradition defined in the main by public service. Despite the fact that it was David Sarnoff, the American broadcaster who first argued that radio 'should be distinctly regarded as a public service' (Briggs 1985: 18), radio in Europe is fundamentally very different to the radio landscape of the United States, where commercial radio and the marketplace rapidly became the dominant force, defining both the radio experience, culture and ethos (Barnouw 1966; Smulyan 1994; Hilliard and Keith 2001; Hilmes and Loviglio 2002; Hilmes 2003). Finally, Europe's radio is also characterised by a long history of being defined and driven by the state, in highly centralized fashion in the case of countries such as France (Meadel 1994), or indeed in former totalitarian regimes of Eastern Europe (Paulu 1974), and along more federal or devolved lines in countries such as Germany, Switzerland and the Netherlands (Kuhn 1985). The development of state broadcasting monopolies in most European countries, established in the early years of the twentieth century following the invention of sound broadcasting, has ensured that there is an enduring shared common ideological approach to radio broadcasting, which now finds expression in the field of digital radio policy.

The roots of European public radio: shaping a shared vision

National state radio stations, broadcasting on MW and LW, began switching on from 1920 and while the initial operating companies were often private businesses, radio broadcasting was quickly seen as a mass communications tool of the state, something to be controlled and regulated (Briggs and Burke 2005: 132). Across Western Europe, state radio evolved into a model of monopoly by public broadcasting, defined and structured in legislation. The arrival of television in the 1950s, and the introduction of commercial broadcasting, particularly in

TV, had the effect of consolidating the dominance of the state-run public service model of radio broadcasting (Tracey 1998). For many European nations, public radio broadcasting remained a monopoly until the 1980s when spectrum was opened to commercial operators, forcing a re-appraisal of radio genres, audiences and markets (Vittet-Philippe et al. 1987).

The dominance of this public broadcasting ethos in Europe, and the governing role that its elites have played, have helped shape a European vision of both the media landscape and its role in society. The post-World War 2 creation of the Common Market and the genesis of the European Union have reinforced this centric view of the broadcasting system and the need for a common European approach to broadcasting in terms of policy and regulation. That ideological view of broadcasting and society as a means of shaping and supporting a shared European philosophy underpins the system as a whole, both public and commercial.

The creation of the European Broadcasting Union (EBU) in 1950 illustrates this concept of Europe as a common broadcasting sphere, with a shared history and culture, driven and defined by the public broadcasting sphere (Fürsich 2004). Formed in the aftermath of World War 2 as an international organization of public service broadcasting (PSB) institutions, the EBU became one of those key European institutions articulating a vision of unification and co-operation as individual countries sought to rebuild their radio broadcasting networks and establish new television services. While individual nation states within Europe framed their audio-visual services and broadcasting systems within national legislation and regulation, the ideal of a shared European space and the quest for a common purpose in broadcasting and society framed its discourse. The collapse of the Soviet Union from the late 1980s, and the emergence of the new East European nation states (Sparks and Reading 1998; Imre 2009), presented a unique challenge since these new states had largely remained in a tightly controlled state broadcasting model and struggled to catch up with the democracy and information matrix of European broadcasting tradition. However, the core European structures of both the European Union and the EBU actively absorbed and assimilated both the new states and the new broadcasters, the EBU formally merging in 1993 with the Organisation Internationale de Radiodiffusion et Television (OIRT), the former organization of Eastern European Broadcasters (Fürsich 2004: 552).

In a quite fundamental sense, technical co-operation in broadcasting has always been linked to this ideological and socio-political cultural framing of radio, through the need for Europeans to co-ordinate and share radio spectrum. That need becomes more crucial the closer you live to your neighbours and for many mainland Europeans (and less so in the case of Great Britain and Ireland) spectrum management is a matter of multi-national diplomacy. International bargaining over the allocation of scarce frequencies was first established with the Geneva plan for European wavelengths in 1926 (Briggs and Burke 2005: 132) and has continued ever since in a series of international conferences to plan, negotiate and manage the radio spectrum.

Likewise, research and development in broadcasting technologies was also part of the EBU's brief from its origins. As a pan-European organisation, embedded with the legacy thinking of the European Economic Community and later the European Union, its impetus was to seek out European technological solutions to ensure that public broadcasters had access to the most advanced technologies available – satellite distribution, digital studio and transmission systems, high definition television, Internet applications – to enable them to maintain a lead against commercial competition, and to support pan-European goals of high technology research and development.

'Similar but different' could be the slogan for European public radio. While the ethos of PSB is encapsulated and defined by the European Commission as a public mandate 'fulfilling the democratic, social, cultural needs of a particular society' (European Commission 2009), Europe has a broad spectrum of PSB models with a wide range of diversity on the economic and social scale from heavy (Norway) to light regulation (Spain), and from zero commercial activity, like the BBC in the UK and Radio France, to small nations like Ireland where PSB is at least 50% dependent on commercial revenue (Van Dijk et al. 2006). The different models of public service broadcasting in Europe reflect the cultural and political characteristics and priorities of different states (Machet et al. 2002). That difference also informs how spectrum has been configured and deployed, creating different radio profiles in terms of stations and audiences. So in Germany, radio reflects the federal Länder structure with regional and local networks, while in Ireland and Portugal local and community radio play a significant role. In some countries, PSB is present at a local, regional and national level (e.g. the BBC in the UK), while in others it is restricted to the national sphere (Ireland), or there is internal national choice between PSB with different operators at national, regional and local levels (the Netherlands). This 'á la carte' PSB map of Europe reflects complex multi-layers of national economic and socio-political as well as cultural differences. It is probably the only way public service broadcasting in Europe could work – by being closely attuned to its harmonisation in terms of technology in spectrum, platforms and receivers in use in each territory.

What the diverse forms of European PSB models share, however, is a socio-political and ethical base which draws heavily on the BBC model shaped by Lord Reith in the 1920s: the concept that public broadcasting is a neutral force in society, using public funds to be independent of both commercial and political bias, and producing content which 'informs, educates and entertains' and which seeks to be universally available (Scannell and Cardiff 1991; Rolland 2005). While diversity exists across the PSB spectrum, the shared ethos, which in Reithean terms incorporates a moral sense of purpose and duty, is expressed also, we argue, in the quest for common technological frameworks and universality of provision for European digital radio.

Eureka 147 – Europe's digital radio solution

The digitization of radio is not uniquely a European issue, but the project of developing a vision and a technology solution for radio in the digital era was very much European in its origins. In order to set the historical context for Europe's digital radio industry, it is useful to examine the broader issues of its origins in European technology development policy, and the distinct socio-political concerns of the late 1980s and early 1990s that helped shape its 'European-ness'. Digital Audio Broadcasting or DAB, also known as Eureka 147, has its origins in the R&D departments of large electronics equipment manufacturers and engineering divisions of broadcasting and telecommunications organizations, as well as various public and private research institutes that constitute Europe's high technology research environment (Lembke 2003). Its development was part of a general effort in the 1980s to develop more efficient transmission systems arising out of the ability to carry information in the form of digital signals. As a relatively recent technology, its history and origins have not been extensively documented, though brief historical surveys are available (see for example: Kozamernik 1995; Gandy 2003; Hoeg and Lauterbach 2003; Rudin 2006; O'Neill 2009).

The DAB project began as a collaboration between Institut für Rundfunktechnik (IRT), the research and development institute for the German broadcasters ARD, ZDF, ORF and SRG/SSR, and the Centre Commun détudes de Télédiffusion et Télécommunication (CCETT), the research institute of France Telecom and TDF. Two essential ingredients of the system were already in development prior to the formal organisation of the Eureka consortium: the audio compression or bit-rate reduction system, pioneered by IRT in Germany, and a new radio frequency modulation system called COFDM, led by CCETT in France. The initial basis of the research was the development of an integrated services digital broadcasting system not specifically dedicated to radio. The DAB bit-stream could in fact be used to transmit all kinds of data including images and slow scan television (Gandy 2003: 3). However, with the crucial support of the EBU and some of the leading broadcasting organisations across Europe (BBC, DR, YLE, SR etc.), a consortium of nineteen organisations from France, Germany, The Netherlands and the United Kingdom was formed in 1986 to develop DAB as a successor for AM and FM radio broadcasting.

The Eureka Project 147 was established in 1987, with funding from the European Commission, to develop a system for the broadcasting of audio and data to fixed, portable or mobile receivers (ETSI 2006). The objectives of the research were to develop 'a European technical standard for Digital Audio Broadcasting' and to seek its adoption as a worldwide standard by international bodies like the European Technical Standard Institute (ETSI) and the International Telecommunications Union (ITU). The technical development envisaged a digitalization of broadcasting distribution, which would produce improved reception compared to FM, particularly mobile reception, and with the potential to offer additional services such as text and other data, conditional access, enhanced traffic services, and picture transmission (Eureka 147 n.d.).

The digital radio system developed under Eureka 147 was a highly successful technical feat of engineering that provided an innovative approach to digital audio and multimedia broadcasting (Hoeg and Lauterbach 2003). It was a highly versatile broadcasting system that could be used for terrestrial and satellite, as well as for hybrid and mixed delivery. For broadcasters it offered a new way of combining multiple audio streams onto a single broadcast frequency called a DAB ensemble or multiplex, thereby making much more efficient use of spectrum. Broadcasters could also vary the number of channels within an ensemble, modifying the bit rates of individual streams. The system is also much more robust with fewer artefacts from interference, noise or channel fading. From the user's point of view, DAB promises more choice, better quality, added text and graphic features and automatic tuning to all stations available.

Planning for digital transmission was conceived on the basis that nationally-based or regionally strong networks (for example, the BBC or a separate multiplex operator) would be primarily responsible for managing the network rather than individual local stations organizing their own transmission. The most cost-efficient coverage was achieved by a network of closely-spaced, relatively low power transmitters, organised into a Single Frequency Network that allowed multiple transmitters to cover an extensive area without mutual interference (Lau and Williams 1992: 12). The greatest spectrum efficiency, therefore, was at the larger national or regional level, and more localized services were much less suited to the system. This bias in transmission was confirmed by the frequency allotments allocated for digital radio broadcasting following the ITU frequency planning meeting in Wiesbaden, Germany in 1995. Many smaller local and community services who had hoped that digital broadcasting would offer more secure access to the mass media market were sorely disappointed to discover that the transmission pattern and licensing structure would not favour their type of radio (as discussed in Chapter Nine. See also Rudin et al. 2004; Lax et al. 2008).

The multiplex organisation of programming, with potentially different providers contributing services, represented a significant reorganisation of the transmission chain (Riley 1994). From the relatively simple structure of broadcaster as content provider and owner of the infrastructure feeding final content into the broadcast chain for transmission and distribution, DAB introduced the distinct functions of programme provider, data service providers and multiplex or ensemble provider (Hoeg and Lauterbach 2003: 152). The DAB configuration thereby required effective co-ordination between each element of the service, and as such was optimally suited to the large broadcaster with the relevant technical and programming resources to serve all aspects of the DAB service. An idealised service provision model, therefore, mapped closely to the kind of programme services envisaged by DAB's main supporters: the large public broadcasting organisations such as BBC, Danmarks Radio or Bayerischer Rundfunk, who had the ability to produce suites of diverse programme material, associated programme data and other listener services under a common brand.

The switching on of the first Digital Audio Broadcasting network in 1995 in the United Kingdom was hailed as a new dawn for radio. It was claimed to be the most significant development since the introduction of FM stereo broadcasting (Bower 1998), and was presented as the replacement technology for AM and FM radio broadcasting (Witherow and Laven 1995). In 1995, with the support of the EBU, the European DAB Forum (EuroDAB) was established to co-ordinate the DAB standard and promote its adoption. This was reconstituted as the World DAB Forum in 1997, representing more than 30 countries, becoming in 2006 the World DMB Forum, which now presides over both the DAB and DMB standard. More recently, variants of the system have included the development of a related digital multimedia broadcasting system or DMB, and a revised and upgraded DAB+ specification using the AAC+ audio codec rather than the original MPEG Layer II, which now provides more capacity and better reliability.

As well as being an inventive technical feat of engineering, DAB was also promoted as the definitive future for Europe's radio in the digital era. Liz Forgan, managing director BBC network radio, described it as a 'historic moment' marking the 'dawn of a third age of radio' – from the original AM mode of broadcasting, which was 100 years old, to FM, over 50 years old, and now into the digital multimedia world of the twenty-first century (Williams 1995). All media – radio, television and the press – it was believed, would adopt digitally-based delivery systems and support convergence between different media platforms (Kozamernik 1995). DAB heralded a new revolutionary era in radio broadcasting (Lambert 1992; Nunn 1995; Witherow and Laven 1995), underpinned by the belief in the necessity for radio to embrace digital technology to survive in an increasingly competitive and complex market. DAB provided the opportunity to keep 'radio not only alive but healthy in an increasingly competitive environment' (Witherow and Laven, 1995: 61). Conversely, radio risked being marginalised if all other broadcasting systems were to go digital and if radio alone remained in an analogue environment (Kozamernik 1999).

Europe's shared vision: aligning digital radio to the Union

The project of Digital Audio Broadcasting emanated from within the European R&D high technology infrastructure, comprising research labs specialising in telecommunications and radio communications research, sponsored by large broadcasting corporations and funded through the inter-governmental Eureka investment programme. Its European origins and world-leading potential were a source of pride to its originators given the intense international competition in electronics. Strengthening the competitiveness of the European audiovisual industry has been a mainstay of European policy since the mid-1980s with an emphasis on the development of a single market, support for regulatory harmonisation and an enhanced, centralised role for the European Commission in the communications sector (Kaitatzi-Whitlock 1996; Levy 1999; Harcourt 2005). The consolidation of the single market in Europe following ratification of the Maastricht Treaty in 1993 led to a wide-ranging set of measures to capitalise on Europe's potential as a global player in communications technologies, and in audiovisual services to rival those of the United States and Japan. The environment thereby created was one with both liberalising market tendencies designed to encourage pan-European broadcasting, as well as interventionist measures to protect cultural diversity and European audiovisual heritage (Collins 1995). Measures such as the *Television without Frontiers Directive* established a thriving single European broadcasting market, particularly for trans-frontier satellite broadcasting, while the MEDIA programme sought to support and protect European audiovisual production and distribution from international competition. One such measure within the MEDIA 92 programme included the venture capital Media Investment Club to support European audiovisual high technology development.

The Eureka programme, which directly supported the DAB research project, was established in 1985 as an inter-governmental initiative to enhance the competitiveness of European industries and to align them more closely with European Union research and development policies. A particular concern of the period was the increasing dominance of the Japanese consumer electronics industry, threatening Philips and other European manufacturers, and support for European technological innovation became a priority (Lembke 2003: 212). A key objective of European investment in technologies like digital radio, mobile communications and satellite navigation systems was to enable standardisation, firstly at the European level and subsequently on global terms in order to create opportunities for world leadership in high technology systems. With regard to digital radio, it was assumed that with the establishment of a common European standard, significant opportunities would be available for European manufacturers to develop a whole range of new products for the consumer electronics and automobile sectors. The development of DAB was frequently characterised as an attempt to emulate the success of GSM, the global standard for mobile phone communication, developed with strong European backing. As a member of the European Commission argued, 'After the digitisation of communications, digital radio is probably, after digital TV, the last chance for Europe to enhance its competitiveness in the consumer electronics sector. […] Europeans who developed the system and invested most in DAB, have to put all their efforts to participate in the exploitation of the system. With such a joint European efforts, DAB can and will repeat the success story of GSM' (Lembke 2003: 214). Initial expectations for DAB as a consumer electronics item were high and market research suggested that Europeans could buy 50 million DAB sets in the first 10 years, with sales then rising to 35 million a year. This compared very favourably with the CD player, which took eight years to achieve annual sales of 5 million (Fox 1994).

A key component of this vision of developing a global standard for digital radio broadcasting was the requirement for public intervention on a pan-European level, with the appropriate political commitment and institutional backing to enable a stable regulatory framework, co-ordination of frequency allocation and a co-ordinated strategic approach to supporting market adoption of the system. The support of the EBU for the DAB project was particularly important in this regard: the EBU had initiated the first series of studies on satellite DAB in the mid-1980s and supported the formation of the consortium for Eureka 147. EBU members were the driving force behind the consortium and the EBU's Technical Department actively participated in its various working groups. Crucially, the EBU as an international organisation provided the essential logistical support in promoting DAB at the International Telecommunications Union and in the preparations prior to the adoption of DAB as an ETSI standard (Kozamernik 1995: 10). The EBU members, the public radio broadcasters, were and continue to be at the forefront of European digital radio services and are its driving force 'from technical testing, to content provision, to marketing and promoting the platform' (EBU 2007: 8). Most importantly, according to the EBU, public broadcasters have been to the fore in promoting the benefits of digitization to citizens, and acting as the social force underpinning the provision of services that commercial broadcasters would be unable or unwilling to do. As such, DAB articulated a vision of sustaining that service into the digital era, embedded with the values of large-scale national and regional broadcasting and incorporated into the architecture of a system that suited PSB needs rather than other forms of broadcasting.

From a European media policy perspective, the focus of attention has more often been on the cinema and television sectors rather than radio, though a central aim of the participating partners in Eureka 147 was to lobby Brussels for an equivalent level of political support for the digital radio sector. As noted earlier, from its inception the ambition of the Eureka consortium was that DAB should be the defining global standard for the digital system to replace analogue AM and FM broadcasting. Within European policy terms, Eureka 147 was the radio industry's vision of its role within communications convergence and the digital revolution. However, the fact of its successful early development and adoption as the first digital broadcasting standard, before rival systems such as Digital Video Broadcasting (DVB), suggested to the Commission that little public intervention would be needed (Liikanen 2001). The subsequent snail's pace of adoption has led to renewed calls for more direct European support. At a European Commission conference in 1998 to examine political support for DAB, World DAB's then chairperson, Michael McEwen, decried the Commission's hands-off approach: 'If it is not led by Europe, then how can you expect the rest of the world to adopt a European technology?'(European Commission 1998).

However, the European policy commitment to removing regulatory barriers, market intervention and the principle of technological neutrality in liberalised communications markets, meant in European Commission terms that success or failure was primarily the responsibility of market players (Liikanen 2001: 4). The counter argument from the radio industry and the EuroDAB lobby group for Eureka 147 was that there was a 'European' dimension to digital radio, i.e. a question of public policy that could only be satisfactorily addressed at a European rather than at a national level, and that diverging regulatory frameworks and implementation strategies in the Member States would lead to fragmentation of the European market. Manufacturers, for example, strenuously argued that the fragmented and disjointed roll-out of digital radio, with successful implementation in some countries and very little in others, was a serious impediment to the development of a new market for digital radio receivers. The prevailing view that radio was a local medium, and the primary responsibility of diverse national and regional authorities, however, worked against any further European co-ordinated action, and as a consequence decisive European Commission support was always qualified.

The guiding assumptions underpinning the development of Eureka 147 DAB were that a robust and mature technology developed within Europe's highly regarded high technology research environment would provide an ideal replacement standard for the international radio broadcasting industry. DAB's version of digital radio built on the proud experience of its trusted and oldest broadcasting institutions and looked confidently to an imagined future in which the major broadcasting institutions would continue to provide more content of higher quality, and in interactive and multimedia formats. It represented an exemplary model of co-operation between European member states, and between public broadcasting organisations and private manufacturers, with the guiding and financial support of agencies such as Eureka and the EBU. Its early achievements in technology design and rapid development of a fully working system suggested that, as hoped, it could indeed become as great a success as GSM had been previously, and would contribute further to Europe's leading role in global technology development. DAB was certainly successful in attaining early international standardization, with adoption of the basic DAB standard by ETSI in 1993 followed by the ITU-R recommendation of DAB as the only digital radio standard in 1994. The allocation of spectrum for terrestrial digital radio broadcasting by the World Administrative Radio Conference (WARC) in 1992 provided a major boost to its international standing, and launched its efforts towards implementation in Western Europe and beyond.

Despite these early promising indicators, DAB deployment stalled and languished in an extended period of early market deployment and adoption with both ongoing successes and reversals. Lacking the sense of urgency and political priority given to analogue switch off for television, radio contended with a multiplicity of delivery mechanisms (analogue and digital) and deferred the question of whether AM and FM broadcasting needed to be replaced. Strategies for the introduction of digital radio have been characterised by a liberal market approach where it is largely left to market forces to decide the fate of particular technologies. As with previous technological developments in the sector, this resulted in long delays in new technology development, competing solutions, confusion for the radio industry and for audiences, and an uncertain environment for future planning. The appeals by the sector that strong market intervention was needed to create an extended single market with harmonised approaches to spectrum planning and market regulation coincided with a shift in policy towards a market for content as opposed to support for particular media such as radio or television. Ultimately, the lack of a 'European' dimension to Eureka 147 and agreement of a need for European-level intervention led to a situation where the development of digital radio has remained a matter for individual member states.

A renewed effort to realign DAB within a vision of common European standards was launched with the announcement by WorldDMB in 2008 of common Digital Radio Receiver Profiles which would act to unify a fragmenting market and build upon the potential for new service launches on the DAB+ platform (Howard 2009). The European digital radio landscape by 2008 had reached a point where there were relatively successful markets for the original version of DAB in the United Kingdom and Denmark; emerging new markets for DAB+ such as Hungary, the Czech Republic and Malta; other countries where DAB and DAB+ coexisted, to ensure an existing listener base was not disenfranchised (e.g. Switzerland and Germany); and countries such as France, which opted for digital multimedia radio based on DMB, or Norway, which adopted a mixture of DAB and DMB combining free-to-air mobile TV with its existing DAB networks and services. As such there has been no single pattern for digital radio across Europe, much fragmentation, no incentive for manufacturers to develop new products or for consumers to have confidence in a common standard. World DMB's solution, to develop a standard specification for receiver profiles that would be capable of decoding DAB, DAB+ and DMB, sought to make the individual country choice less of an issue. The goal of a digital radio without frontiers therefore is to ensure manufacturers benefit from a single European market for digital radio receivers while minimising the complications arising from different strategies and approaches to digital radio adoption within the Eureka 147 family of standards.

Back to the future: the failing quest

The transition to digital radio has echoes of an earlier technology leap for radio when FM superseded AM in the 1970s as the preferred mode of transmission – see Chapter Three. This, combined with transistor radio and the political upheaval of the age, placed radio at the centre of a social and popular movement, closely connected with the boom in the music industry at the time (Barnard 1990; Negus 1993; Neer 2001). But while that was a zeitgeist moment which allowed radio to be liberated and become mobile, it is almost the opposite of terrestrial digital radio's painfully slow journey from 1995 onwards. The FM transition was global, universal, ensuring that all radios were sold with the same choices and allowing listeners to tune with ease between AM and FM, while the 1990s saw global radio divide, based on geographical need, and offering competing digital futures for radio. The future scenario for digital radio, at least from the late 1990s on, depicted a future where you would potentially need different radio sets when you were travelling, particularly the US which was already firmly outside the DAB experiment by that stage.

Central to understanding the European digital radio story is that Eureka 147 and the DAB project were seeking to answer the question of digital migration and spectrum management on the basis of national and spectrum structures and interests that pertained in Europe's 1980s media landscape. A clear motivation for terrestrial digital within individual countries was lack of spectrum, and as one of the core issues for the European roll out of DAB was that while some countries like the UK had hit spectrum scarcity, others still had as much as 25% of FM sitting empty (Ox Consultants 2004). The origins of European radio, in centric-focussed national structures under a European umbrella, required a terrestrial solution to best serve national and perceived geo-political interests. The counter development of satellite and Internet radio from the late 1990s on were challenging in that they presented global media solutions which threatened to undermine or de-stabilise the status quo, and which appeared to suit commercial interests over public media (Kozamernik et al. 2002).

While the European radio landscape has now evolved, and its structure, audiences and content are shared with commercial radio, the technological innovation of radio continues to be framed by some of the founding assumptions of the Eureka DAB project and led by the combined interests of national and European media policy interests. In many ways the development, roll-out and implementation of terrestrial digital radio in Europe has been defined by this need to serve not end users and markets, but national and European interests, organised around a concept of 'European-ness' and underpinned by a shared vision and common purpose. This dichotomy helps frame our understanding of digital radio, specifically DAB technology. It also helps to explain the slow pace of its adoption and some of the difficulties it later encountered connecting with both users and with the marketplace. The quest was never to find the best or most innovative neutral digital radio solution – it was, by the very nature of who framed the question, to find the best and most innovative digital radio solution which would best serve the needs of the status quo – in this case the concept of European unity and ideology as led by the EBU and its network of public broadcasters.

DAB envisaged a future very much within the boundaries and limitations of the present. It was proposed as a technology that could replace analogue radio and its requirements for stationary and mobile reception, and which would best protect, and even replicate, the existing national and supra-national interests. Those interests, as defined by the key institutions governing Europe's radio, in many ways shaped the blinkers of how the technology was envisaged and implemented.

While the concept of a common digital future with one single platform replacing analogue radio may still seem distant and confused, it remains a key motivator for key institutions within the European centre. The prospect of digital diversity, with each country mapping its own path using a variety of platforms whether DAB, DAB+, DVB, or DRM, is seen as one which will lead to radio fragmentation and a weak market (Howard 2009). In the midst of digital divergence, the global, economic slow-down since mid-2008 has hit media badly, with advertisement and sponsorship falling. In the UK, the digital radio market received a significant setback in October 2008 when Channel 4 cancelled its digital radio plans due to the changed economic environment. Since then, many UK commercial digital radio stations have been switched off because of falling revenue.

The poor financial climate for radio has seen dramatic cuts in radio operations in both commercial and public radio. While the commercial sector has been hit by falling advertising revenue, many public broadcasters have also experienced reductions in state revenues. The potential to invest and grow digital radio networks and original channels of content is low, and at the same time the role of the Internet as a global content platform for both audio and video is growing.

Yet the need for a shared vision on radio's digital future, the need for a common platform, remains. As Europe both enlarges its membership and moves closer in its institutional decision-making, the integration of digital terrestrial platforms and receivers in television and radio will promote a more effective union. While the United States equally struggles with digital radio platforms, the compelling logic is for a harmonised approach. Ironically, the Internet remains the common base; the global platform that increasingly sees all forms of content converge and flow. While it is currently not a replacement for terrestrial analogue services, the future may yet be defined more by its successes than by digital terrestrial's failures. The BBC's proposed RadioPlus player effectively takes the best of both the online and terrestrial offerings and presents a user-focused radio with on-demand interactivity, re-thinking radio's proposition to take full advantage of digital technology (Martinson 2009). That marriage of a DAB+ network with high speed Internet connectivity, providing the best of linear and non linear radio, podcasts and interactive tools (like search, archiving and programming), may be the beginning of radio's re-invention. What has become apparent in the past fifteen years is that the simple transition from analogue to digital terrestrial networks is not sufficient; radio needs to re-think itself from the inside out. The European vision may have seen radio's future as secured by the unity of the Eureka 147 project, but, in reality, radio's future may more likely be based in a multi-platform, multimedia base that recognises the changed relationship between content and the audience, and between content and the market.

References

Barnard, S. (1990), *On the Radio: Music Radio in Britain*, Milton Keynes: Open University Press.

Barnouw, E. (1966), *A history of broadcasting in the United States*, New York: Oxford University Press.

Bower, A. J. (1998), 'Digital Radio – The Eureka 147 DAB System', *Electronic Engineering*, April 1998, pp. 55–56.

Briggs, A. (1985), *The BBC: The First Fifty Years*, Oxford: Oxford University Press.

Briggs, A. and P. Burke (2005), *A social history of the media : from Gutenberg to the Internet*, Cambridge: Polity.

Collins, R. (1995), 'Reflections across the Atlantic: Contrasts and Complementarities in Broadcasting Policy in Canada and the European Community in the 1990s', *Canadian Journal of Communication*, 20:4.

EBU (2007), *Public Radio in Europe 2007*, Geneva: European Broadcasting Union.

Eureka 147. (n.d.), 'Eureka Project 147- DAB (Imp)', http://www.eureka.be.html. Accessed 1 June 2009.

European Commission (1998), 'Radio in the digital era: A Report on the Meeting organised by the European Commission (DG X)', Brussels: European Commission (DG X).

European Commission (2009), 'Draft Communication From The Commission On The Application Of State Aid Rules To Public Service Broadcasting Brussels', Brussels: Commission of the European Communities.

European Telecommunications Standards Institute (2006), 'Radio Broadcasting Systems: Digital Audio Broadcasting (DAB) to mobile, portable and fixed receivers', Sophia Antipolis Cedex.

Fox, B. (1994), 'The perfect sound machine', *The Times*, London, 14 October 1994.

Fürsich, E. (2004), 'European Broadcasting Union', in C. Sterling and M. Keith (eds.), *Museum of Broadcast Communications Encyclopedia of Radio*, 1, New York: Routledge, pp. 552–555.

Gandy, C. (2003), 'DAB: an introduction to the Eureka DAB system and a guide to how it works', *BBC R&D White Paper*. WHP 061.

Harcourt, A. (2005), *The European Union and the regulation of media markets,* Manchester: Manchester University Press.

Hilliard, R. L. and M. C. Keith (2001), *The broadcast century and beyond : a biography of American broadcasting,* Boston, MA. and Oxford: Focal Press.

Hilmes, M. (2003), 'British quality, American chaos', *Radio Journal: International Studies in Broadcast & Audio Media,* 1:1, pp. 2–17.

Hilmes, M. and J. Loviglio (2002), *Radio Reader: Essays in the cultural history of radio,* London: Routledge.

Hoeg, W. and T. Lauterbach (2003), *Digital Audio Broadcasting Principles and Applications of Digital Radio,* Chichester: John Wiley & Sons.

Howard, Q. (2009), 'One digital radio across Europe', *EBU Technical Review,* pp. 1–11.

Imre, A. (2009), *Identity Games: Globalization and the Transformation of Media Cultures in the New Europe,* Boston, MA: MIT Press.

Kaitatzi-Whitlock, S. (1996), 'Pluralism and Media Concentration in Europe: Media Policy as Industrial Policy', *European Journal of Communication,* 11: 4, pp. 453–483.

Kozamernik, F. (1995), 'Digital Audio Broadcasting – radio now and for the future', *EBU Technical Review,* 265.

Kozamernik, F. (1995), 'Eureka 147 to a worldwide standard', Presentation at *Audio Engineering Society. DAB – The Future of Radio,* London.

Kozamernik, F. (1999), 'Digital Audio Broadcasting – coming out of the tunnel', *EBU Technical Review,* 279.

Kozamernik, F., N. Laflin and T. O'Leary (2002), 'Satellite DSB systems – and their potential impact on the planning of terrestrial DAB services in Europe', *EBU Technical Review,* 289.

Kuhn, R. (1985), *Broadcasting and politics in Western Europe,* London: Cass.

Lambert, P. (1992), 'DAB: Signs of coming down to Earth', *Broadcasting,* 122: 25, pp. 37–1/2.

Lau, A. and W. F. Williams (1992), 'Service planning for terrestrial Digital Audio Broadcasting', *EBU Technical Review,* 252.

Lax, S., M. Ala-Fossi, P. Jauert and H. Shaw (2008), 'DAB: the future of radio? The development of digital radio in four European countries', *Media Culture Society,* 30:2, pp. 151–166.

Lembke, J. (2003), *Competition for Technological Leadership: EU Policy for High Technology,* Cheltenham: Edward Elgar Publishing.

Levy, D. A. (1999), *Europe's digital revolution broadcasting regulation, the EU, and the nation state,* New York: Routledge.

Liikanen, E. (2001), 'Prospects for Digital Radio Development in the European Union', Presentation at *World DAB Annual General Assembly,* Brussels, 8 November 2001.

Machet, E., E. Pertzinidou and D. Ward (2002), 'A Comparative Analysis of Television Programming Regulation in Seven European Countries: A Benchmark Study', Report by the *European Institute for the Media* commissioned by the NOS, Düsseldorf: European Institute for the Media (EIM).

Martinson, J. (2009), 'BBC bids for online catchup service for UK's radio stations', *The Guardian,* 23 March 2009.

McDonnell, J. (1991), *Public Service Broadcasting: A reader,* London: Routledge.

Meadel, C. (1994), 'Between corporatism and representation: The birth of a public radio service in France', *Media, Culture & Society,* 16:4, pp. 609–24.

Neer, R. (2001), *FM: The Rise and Fall of Rock Radio,* New York: Villard.

Negus, K. (1993), 'Plugging and programming: pop radio and record promotion in Britain and the US', *Popular Music,* 12, pp. 57–68.

Nunn, J. (1995), 'Introduction', Presentation at *Audio Engineering Society. DAB – The Future of Radio*, London.

O'Neill, B. (2009), 'DAB Eureka 147: a European vision for digital radio', *New Media Society*, 11: 1–2, pp. 261–278.

Ox Consultants (2004), *Final Report: Review of Licensing of Radio Services in Ireland. A Study for the Department of Communication, Marine and Natural Resources*, Dublin: DCMNR.

Paulu, B. (1967), *Radio and television broadcasting on the European Continent*, Minneapolis: University of Minnesota Press.

Paulu, B. (1974), *Radio and television broadcasting in Eastern Europe*, Minneapolis and London: University of Minnesota Press.

Riley, J. (1994), 'DAB multiplex and system support features', *EBU Technical Review*, 259.

Rolland, A. (2005), 'Establishing Public Broadcasting Monopolies: Reappraising the British and Norwegian Cases', *Media History Monographs*, 8:1, pp. 1–22.

Rudin, R. (2006), 'The Development of DAB Digital Radio in the UK: The Battle for Control of a New Technology in an Old Medium', *Convergence*, 12:2, pp. 163–178.

Rudin, R., W. A. K. Huff, G. F. Lowe and G. Mytton (2004), 'Digital Audio Broadcasting', in C. Sterling and M. Keith (eds.), *Museum of Broadcast Communications Encyclopedia of Radio*, 1, New York: Routledge, pp. 456–462.

Scannell, P. and D. Cardiff (1991), *A Social History of British Broadcasting, Vol.1 1922–1939*, Oxford: Basil Blackwell.

Smulyan, S. (1994), *Selling Radio: The Commercialization of American Broadcasting, 1920–1934*, Washington, D.C: Smithsonian Institution Press.

Sparks, C. and A. Reading (1998), *Communism, capitalism and the mass media*, London: Sage.

Tracey, M. (1998), *The decline and fall of public service broadcasting*, Oxford: Oxford University Press.

Van Dijk, M., R. Nahuis and D. Waagmeester (2006), 'Does Public Service Broadcasting Serve The Public? The Future of Television in the Changing Media Landscape', *De Economist*, 154: 2, pp. 251–276.

Vittet-Philippe, P. and P. Crookes (1987), *Local radio and regional development in Europe*, Manchester: European Institute for the Media in association with the European Centre for Political Studies.

Williams, R. (1995), 'BBC switches on CD-quality radio', *The Independent*, London, 28 September 1995.

Witherow, D. M. L. and P. A. Laven (1995), 'Digital audio broadcasting-the future of radio', *International Broadcasting Convention, IBC 1995*, Amsterdam, Netherlands.

Chapter 2

The Technological Landscape of Radio[1]

Marko Ala-Fossi

The future of radio is now much less obvious and clear than it was in the mid-1990s. Instead of a relatively straightforward transition from analogue to digital audio broadcasting (DAB), and increasing convergence as a result of digitalization, we have rather witnessed divergence, which has resulted in a wide selection of both competing alternative and complementary technological options for digital audio delivery. Some have similarities with analogue radio broadcasting systems; others are challenging the idea of radio as solely an aural medium, or even the concept of broadcasting with its tradition of linear production and reception, all of which are embedded into the traditional definition of radio. Consequently, depending on one's perspective, radio is currently either facing the danger of fragmentation or is surviving by infiltrating new platforms and becoming more polymorphic.

This chapter intends to be a roadmap through the jungle of cryptic names and abbreviations of the systems currently forming the new technological landscape of radio. It aims to provide a socio-economic analysis and comparison of these different technologies using the perspectives of political economy and media economics (Mosco 1996; Picard 1989), as well as technology assessment (Mäkinen and Nokelainen 2003; Henderson and Clark 1990) and diffusion of innovations theory (Rogers 2003). In addition, this historical overview also examines and describes the underlying reasons why and by whom these different and diverse systems have been developed, and the way that they have been developed (Mackay and Gillespie 1992; MacKenzie and Wacjman 1999).

Towards digitalization – and convergence?

By the end of the 1930s, FM radio and television had been introduced in the United States and AM radio broadcasting had already become a mass medium, but the technological development of analogue broadcasting was not over yet. New ways to make the broadcast media experience more realistic and natural, such as FM stereo and colour television, were still to be introduced to the consumers of the 1960s. However, the first signs of development towards digital broadcasting and digital radio are to be found in the early 1970s when broadcast engineers, at least in Europe, began to study the possibilities of improving and enhancing analogue broadcasting with digital technology.

Early digital broadcast innovations made their first appearance at this time. Teletext, a service originally designed to provide television subtitles for the deaf, was created at the BBC and went live in mid-1970s. It is not purely coincidental, perhaps, that the pre-development of Radio Data System (RDS) was started in the European Broadcasting Union (EBU) at the same time, although the system was not introduced until late 1980s (BBC 2004a, RDS 2008). The basic idea of RDS is quite similar to Teletext, in that in both cases the extra capacity of the analogue system is used to break the broadcast flow and provide additional digital services for an audience with suitable broadcast receivers. Both Teletext and RDS fit well with a public service ideology, as well as with the strictly separate radio and television divisions of contemporary European public service broadcasting organizations. These new textual dimensions did not challenge established boundaries between the two broadcast media, but they did provide new information services for the public, free of charge.

At the same time, Sony and other leading Japanese electronics companies were facing increasing competition from cheaper East Asian manufacturers and accordingly decided to develop a new analogue standard for High Definition Television (HDTV) so that the Japanese industry could concentrate on production of more advanced devices. The research and development work inside the Japanese public service broadcaster NHK took some time, but in 1985 Japan (with the United States and Canada) proposed its HDTV system as the new world standard. European countries interpreted this as an attempt to take over the world market, and as a consequence, proposed a different approach. With the support of the EU, they also launched their own HDTV development project [Eureka 95] in July 1986 with leading European electronics companies including Philips, Thomson and Bosch (Schiffer 1991; Wood 1995; Soramäki and Okkonen 1999).

The development project of the Digital Audio Broadcasting (DAB) system [Eureka 147] launched just six months later and, with the central involvement of the same three companies, can be seen as another part of the European counter-attack. This was reflected in the original Eureka project form, which states that 'the drawing up of a new digital audio broadcasting standard will therefore provide a long term counterbalance to the increasing dominance of the countries of the Far East in the consumer electronics sector' (Eureka 1986). As noted in Chapter One, the project group also estimated that there would be a good chance to get the new DAB system adopted worldwide to give the European electronics industry a competitive advantage (Rudin 2006; Lax 2003) – partly because the Japanese had concentrated more on HDTV than on digital radio.

However, it is unlikely that the Japanese would have been interested in developing a digital radio system in the first place. In Japan, radio is considered mainly as a small and secondary medium for car drivers or as an accompaniment to household chores (Kato 1998). Moreover, by 1982 NHK researchers had already drafted a *digital highway* concept, later known as Integrated Services Digital Broadcasting (ISDB). The idea was a single digital broadcasting system, which could be used for delivering all kinds of information in digital format (Yoshino 2000). Interestingly, the concept of ISDB was also mentioned in the Eureka 147 project form (Eureka 1986), so the project participants were obviously aware of the Japanese idea to increase *network convergence,*[2] but they might not have understood the idea or even have agreed with it. By contrast, European engineers and later the EBU and its members were focused on *network digitalization*, crafting separate media-specific digital systems, first for radio and then for television, instead of integrated solutions.

As noted by O'Neill and Shaw (this volume), the Eureka 147 project did not start from scratch. European pre-development of DAB began already in 1981 at the Institut für Rundfunktechnik (IRT), obviously inspired by the Compact Disc (CD) standard, which had been introduced in 1980 (Immonen 1999; Immink 1998). The original focus in the early to mid-1980s was to improve radio by developing a digital audio broadcasting system to make the most of new CD recordings with crystal-clear sound, as well as new RDS features in mobile reception and to replace FM (this is discussed further in Chapters Three and Four). But just improving radio was not enough for all the project partners; they wanted to extend it. Although the idea of including data transmission and multimedia capabilities in DAB may seem an afterthought of the 1990s, it was briefly mentioned in the Eureka 147 project form as part of the original concept (Eureka 1986; Gandy 2003).

The further development of DAB into an operational broadcasting system took quite a few years[3] during which the surrounding societies were going through several important social, political and economic changes, as well as a major technological transition in communications and media. The foundations of DAB design had been laid before the wider introduction of private and local broadcasting in Europe in the late 1980s, while the early 1990s was a turning point in development of mobile telephony and the Internet. Mobile phones had already been available for a while, but the European digital GSM (Global System for Mobile communications) standard, commercially introduced in 1991, turned out to be highly successful: within 10 years, there were 500 million GSM subscribers worldwide (GSM World 2008). In similar fashion, the Internet was gaining strong momentum in the late 1980s, though it did not have any serious impact on traditional media before the introduction of the World Wide Web in 1992 and the first web browser in 1993 (Henten and Tadayoni 2008). This was also the moment when concepts such as 'convergence' and 'information society' became key issues in the trade and broadcast industry press as well as in communications research in the Western world. All this had little immediate impact – if any – on research and development of digital broadcasting systems in Europe or in the US.

So radical – but yet conservative

As noted in Chapter One, when the DAB system was introduced in mid-1990s in Europe, it was expected to be a comparable transition for radio as the transition to digital GSM had been for mobile phones a few years earlier. There was an optimistic, perhaps even deterministic, assumption that the superior new digital radio system would replace analogue FM radio in a relatively short time in smooth evolution (Kozamernik 1995; Mykkänen 1995).

DAB was technologically radical, based on new innovations which were also combined in a new way in order to extend the limits of broadcast radio. It was designed for not only delivering sound, but any sort of data (text and pictures, even video clips or web pages) with good mobile reception. In addition, the system was allocated partly new frequencies outside existing broadcasting bands. One DAB transmitter in a Single Frequency Network (SFN) could broadcast several programmes at the same time, with programme services combined into a multiplex. With this new system it was possible to deliver five CD-quality audio services on the same frequency. The fact that some twenty years later most digital broadcasting systems are still using the same OFDM modulation system originally applied in DAB, reflects the enduring quality of engineering work – though later advances in audio coding technology made its original (MPEG-1 Audio Layer II) audio coding relatively inefficient by today's standards (Lax 2003; Kozamernik 2004).

Although DAB was a radical innovation from a technological perspective, it was rather conservative from a social point of view, reflecting the needs and wants of the national PSB radio organizations that dominated European radio broadcasting before the introduction of private local radio. The basic design of DAB fits better with large nationwide broadcasting systems and, consequently, it has been more successful in countries with strong public service radio organizations (Hendy 2000). Interestingly, the same characteristics of DAB that made it socially conservative in Europe made DAB radical in social as well as economic perspectives in the U.S., due to the different social relations and economic structures of American society. Although the National Association of Broadcasters (NAB) had originally recommended the adoption of DAB, the U.S. commercial radio industry saw it as a threat to the status quo and started in 1991 to develop an alternative digital radio system for already allocated AM and FM radio wavebands (Ala-Fossi and Stavitsky 2003). This was DAB's major set-back: from this point on DAB lost its chance of becoming a worldwide system for digital radio.

The primary expectation was that the new DAB system would provide a better radio service and would also enhance radio with new text, data and multimedia services – while some believed that data-DAB could become even more important than digital radio (Mykkänen 1995). In Finland, mobile phone giant Nokia started to study the multimedia broadcasting capabilities of DAB in 1995, but withdrew prematurely from the project and ceased all DAB development in 1997 (Kuusela 1996; Kuusisto 1998). Others were more persistent. The BBC, for example, started to work on DAB text services in 1996, a year after the introduction of DAB in the UK, and piloted data enhancements in 1998 (BBC 2004b). The Bosch Group continued their own work with DAB multimedia, and before the end of decade they had developed a DAB-based system called Digital Multimedia Broadcasting (DMB) which could also deliver video for mobile reception (Kozamernik 2004).

By that time the most optimistic radio managers and receiver manufacturers already thought the term 'digital radio' was a misnomer because the range of possible new wireless DAB multimedia services was so wide (Beatty 1998). Indeed, some radio researchers were considering whether DAB with its new visual dimension would be radio at all (Åberg 2001). However, in the age of the Internet, the multimedia capacity of DAB and the usefulness of these new services turned out to be relatively modest (Mäntylä 2000). Moreover, most of the rather expensive first-generation DAB receivers were not even capable of receiving multimedia services or offer much more than an analogue FM receiver with a RDS tuner (Kozamernik 2004; Lax 2003). Bosch did some DMB experiments with the public transportation systems in Germany (Tunze 2005), but interest in this multimedia system remained very low.

According to one estimate, €500 million had been spent on DAB development in Europe by 1999 (Kozamernik 1999). As Chapter One describes, despite the early support and great hopes for international success, the EU was not particularly active or generous in supporting the development of DAB or digital radio. However, with the support of the World DAB organization and the EBU, it was widely adopted in Europe and altogether in 28 countries around the world within ten years of its introduction (World DAB 2005). But even in those countries where DAB was implemented, the adoption rate of the new radio system was low: consumers were not interested in buying new, rather expensive digital receivers as eagerly as expected and the early phase of the classic S-shaped diffusion curve (Rogers 2003) was much less steep than anticipated. Despite the promises of new features, more choice and better sound, the benefits of DAB radio over FM radio were not obvious to everybody (Lax 2003). The largest DAB market in the world with over seven million receivers is the United Kingdom, but the share of radio listening via DAB even there is only 11.4% (World DMB 2009, Plunkett 2009a).

DVB and DRM – supplementing DAB

Because analogue HDTV had not been able to make a breakthrough and DAB was intended as the digital future of FM radio, digitalization of television needed its own pan-European project. The preparations for Digital Video Broadcasting (DVB) were started in 1991 and the project was formally launched two years later. The digital television system was similarly going to be radical with certain technical similarities to DAB such as multiplexing and the Single Frequency Network system, but it had also fundamental and important differences which reflect how the designers had interpreted and understood the idea of television. The new DVB system was designed primarily for stable roof-top antenna reception of video content on wideband channels. As a result, when DVB-T (Terrestrial) was approved as a standard in the 1995, it was inappropriate for any serious mobile use, and also uneconomic for large-scale delivery of radio-type services (Wood 1995; Wood 2001; Laven 1998; Kozamernik 1999; DVB 2008a). In practice, the new European digital broadcasting systems for radio and television were less compatible with each other than the analogue broadcasting systems (both using FM for sound) had been. Despite this, the existing DVB-T networks have been used for radio services in quite a few European countries (like UK and Finland) alongside or instead of DAB.

After FM radio and television, another European project was needed to bring AM broadcasting into the digital age. A group of broadcasters and manufacturers formed the Digital Radio Mondiale (DRM) consortium in 1998 to develop a digital system for broadcasting bands below 30MHz, so-called AM bands. DRM is a modular innovation similar in approach to the strategy preferred by the American radio industry.[4] It is designed to use existing AM channel allocations – or only half of the channel bandwidth. DRM is capable of providing near-FM quality sound – an audible improvement for AM broadcasting – and also has the capacity to broadcast additional data and text. The DRM system uses the same modulation technique as DAB, but with three more up-to-date options for audio coding (including MPEG-4 AAC) (Olon 2002; DRM 2005b).

DRM is obviously much closer to the idea of digitalization of radio media in its traditional form than DAB. Although DRM is in a technological sense at least as advanced a system as DAB, from a social and economic perspective it is much less radical and demanding. DRM has less capacity for data and sound, and does not challenge the existing economic and social structures of broadcasting. With DRM, each broadcaster can keep its own transmitters and networks and continue operations on separate frequencies instead of joining a multiplex. No wonder that in 2005 the World DAB Forum and DRM consortium set up a joint project to enhance DRM for the FM band also (DRM+) (Hallet 2005; DRM 2005a).

There is no particular technological reason why DRM has emerged as an alternative replacement of FM broadcasting. It can be argued that the main reasons are actually social and economic. DAB turned out not to be a 'one-size-fits all' solution; although it is well suited for nationwide broadcasting, it is a rather expensive and even cumbersome technology for small-scale FM broadcasting services such as community radio (Olon 2002; Coonan 2005). On the other hand, DRM on the AM dial can be used for covering very large areas – even across national borders – with fewer transmitters. In December 2008, the BBC and Deutsche Welle launched a new joint DRM radio channel covering most of Western Europe (France, Germany, Belgium, Netherlands and Luxembourg) with only six transmitters (DRM 2008c).

The new DRM+ system is said to be suitable for small broadcasters with a single station on a single frequency with localised coverage, as well as for large broadcasters with huge coverage areas and multiple channels on a single frequency network. The test results so far have been very positive and the new system was accepted as a digital broadcasting standard in 2009. According to the DRM consortium, DRM+ will be important as a replacement technology for FM, especially for regional and local broadcasters, whether there are existing DAB networks or not (DRM 2008a, DRM 2008b). Together, it would appear that these two European systems could provide compatible and tailor-made solutions for digital radio broadcasting, leaving rival technologies with much less chance to enter the market.

Sounds from the sky

The first digital satellite radio system was developed in the mid-1990s by French Alcatel for a U.S-based satellite radio company, WorldSpace, which launched two geo-stationary satellites to cover Africa (1998) and Asia (2000). The WorldSpace satellite system, which is expected to remain functional for 15 years, can deliver up to 200 audio programs per beam as well as some multimedia content (Alcatel Space 2001). In 2001, XM Satellite Radio launched its services in the United States with two geo-stationary satellites over North America, using the same satellite technology as WorldSpace and offering up to 100 channels. In 2002, another U.S. satellite radio service called Sirius launched three geo-synchronous satellites on elliptical orbits to deliver up to 100 channels. All three systems have different audio coding and each of them needs a dedicated receiver and a subscription (XM 2000; Layer 2001). There are also other less used technologies for delivering audio content via satellite, such as S-DAB (satellite version of DAB) and DVB-S, which means that audio services can be received with digital satellite television. For example, SES ASTRA currently offers in Europe almost 350 digital radio services on DVB-S (ASTRA 2009). In addition, there is the satellite version of DMB, the so-called S-DMB. This satellite-based system for mobile multimedia services, based on S-DMB technology, was developed and is owned by the Toshiba Corporation. It launched in Japan and South Korea in 2005 (Suk 2005; Furukawa 2005).

Digital satellite radio broadcasting is radical when compared to terrestrial (analogue) broadcasting. It can easily cover a continent and offer hundreds of options simultaneously for potentially millions of people. It is also a very capital intensive approach, requiring huge initial investment not only in satellite systems but also in terrestrial repeater networks. In addition, the U.S. satellite radio companies had to pay substantial sums (Sirius $89 million, XM $92 million) for spectrum space (Frenzel 2003). The concept of offering uninterrupted and clear coverage from coast to coast with a variety of channels and very few or no advertisements for a premium price has been a success in the U.S. – in terms of the number of paying subscribers (almost 20 million).[5] However, both XM and Sirius have been unable to make a profit, and the former rivals were allowed to merge in 2008 (Arango 2008). Even so, the combined Sirius XM has a huge debt load – and two sets of satellites covering the same area.

Europe, with its many languages and plentiful supply of advertisement-free radio programming, is perhaps a much more difficult market than North America, but at least three different groups have recently been planning to launch dedicated satellite radio services for European audiences. One of them, Madrid-based Ondas Media, still plans a full European launch in 2012 (Krueger 2009). In addition to commercial projects, the European Space Agency (ESA) has introduced a new satellite multimedia car radio system designed to use already existing, old TV satellites and to work without a terrestrial repeater network (ESA 2007).

Digital terrestrial radio – the American way

The American approach to digital radio was originally called IBOC (In-Band, On Channel) because its starting point was to design a digital radio system for existing broadcasting bands and present channel allocations. The task was not easy. After rejecting DAB, competing development projects during the 1990s made at least nineteen different proposals for IBOC or IBAC (In-Band, on Adjacent Channel), all of which failed. By the end of the decade, IBOC developers began joining forces and in July 2000 they merged into a single development company, iBiquity Digital Corporation. Time was running out because the satellite radio companies were about to launch their new nationwide digital radio subscription services. iBiquity was finally able to introduce an operational, but not a complete, version of IBOC in April 2002. A couple of months later it was approved as the technical standard for terrestrial digital radio broadcasting in the United States (Ala-Fossi and Stavitsky 2003).

IBOC was very carefully designed to be a modular innovation in relation to analogue broadcasting. An FM signal needs only about half of the 200 kHz channel bandwidth available in the United States. By using the so called sidebands on both sides of the analogue signal for the digital signal, it is possible to simultaneously broadcast a standard analogue signal plus one near CD-quality digital signal and a small amount of additional data. Later, the bandwidth of the analogue signal could be reduced and, in the end, completely replaced with a third digital signal. The reason for this alternative approach is primarily economic. Because IBOC stations use practically the same spectrum space as before, there is no need for a new licensing round; there is no risk of new operators, uneven sharing of the new frequency spectrum or an expensive license auction. And because the most valuable assets of radio corporations are usually their broadcast licenses, the most fundamental reason for developing IBOC technology was that the US broadcast industry wanted to protect the asset values of their current broadcast licenses and corporate stock (Ala-Fossi and Stavitsky 2003).

IBOC – currently marketed as HD Radio[6] – is an attempt to fight the digital challenges of satellite and Internet radio and, at the same time, an attempt to preserve the existing economic structures and power relations of US broadcasting industry. Because the system uses current analogue frequencies already licensed to incumbent broadcasters, even the new multicast radio channels are run by the same operators. It will not provide any spectrum recovery. In practice, HD Radio will leave even less spectrum space for new operators or services. Unlike European standards, HD Radio is a proprietary system: every operator broadcasting with the system has to pay a fee to iBiquity (Ala-Fossi and Stavitsky 2003).

Although the radio industry has claimed that thanks to its different approach HD Radio would provide a more 'seamless' transition from analogue to digital radio than DAB, the use of services on either of these digital platforms requires people to buy new receivers. Six years after its introduction, HD radio has not been able to make a real breakthrough, even in the United States. In 2008, there were only about 1828 HD stations on air and the number of the new stations switching to HD was declining already for a second year in a row (PEJ 2009).

Sayonara to single media networks

Although the concept of ISDB was invented in the 1980s, the technical specification for this kind of digital broadcasting system did not exist until 1997. ISDB-T (T for terrestrial use) has several technical commonalities with DAB and DVB, and it would appear that Japanese engineers utilized a lot of European research. Kim (2003), for example, claims that ISDB-T is modified from Eureka DAB and is more or less a DAB-derivative (Yokohata 2007; Nakahara 2003; Miyazawa 2004). Because ISDB is designed to be a fully scalable system, all free digital television and radio channels on ISDB, with their additional digital services, can also be received with a small handheld mobile (television) receiver (Ikeda 2006). The so called 'One-Seg' mobile television service was launched in 2006, and a year later about ten million handheld receivers were sold in Japan (Valaskivi 2007; Takahara 2006; Yokohata 2007). In 2008, the number of users had already doubled to twenty million (O'Brien 2008).

ISDB may appear to be a combination of already existing technologies, but its approach is radical because it merges all broadcast networks into one converged platform for all kinds of media content. As a consequence, the broadcast media content which the consumer is able to receive is mostly dependent on the type and capacity of the end user terminal or receiver. On an ISDB network, the traditional differences between audiovisual media and audio media are no longer obvious. Moreover, this approach also provides opportunities to explore the uncharted digital borderland between traditional broadcast radio and television – by developing new concepts like *visual radio* or *mobile television*. The development of ISDB seems to have also had economic motives: it protects the Japanese electronics industry from foreign competition. In addition, if other countries happen to choose ISDB-T as their digital broadcasting standard, this of course would benefit Japanese manufacturers. Since 1997, the developer of the system, the Association of Radio Industries and Businesses (ARIB), has promoted ISDB-T and ISDB-TSB (SB for Sound Broadcasting) internationally, but so far only Brazil has adopted ISDB-T for its digital terrestrial television standard (Miyazawa 2004; DVB 2008b).

Web radio and Internet audio – with and without wires

Delivery of digital audio files over the Internet had became possible in the early 1990s, soon after the introduction of the World Wide Web and browsers, partly thanks to the new and efficient audio coding system of mp3 (MPEG-1 Audio Layer III), developed at the German Fraunhofer Institute simultaneously with audio coding for DAB (Fraunhofer 2009). First attempts to stream audio over the Internet were also made by that time, and after the introduction of the essential computer software such as Real Audio, streaming started to gain popularity (Priestman 2002; Menduni 2007). Earlier when one had to first download the audio file and then listen, the stream player made it possible to listen to the audio while also receiving it. The sound quality was not very high at first, but the number of radio stations offering audio streaming services over the Internet grew rapidly in the 1990s, especially in the United States (Lax and Ala-Fossi 2008).

Web radio is an architectural innovation that mainly combined already existing elements. Unlike the newly developed European digital broadcasting systems, the Internet was truly able to deliver all kinds of digital content in one single network – for example, pictures and text on an audio streaming website. In addition, with the Internet there are no equivalent regulations or shortage of frequencies as in broadcasting, and the services have the potential to reach a global audience. Web radio also has certain socio-economic strengths because it is possible to create services for very small and scattered groups without very large initial investments. No wonder that the Internet has become the main delivery system for thousands of web-only radio operators and an important supplement for practically all (radio) broadcasters. However, there are reasons why Internet audio services are more likely to complement broadcast radio than substitute it.

The increase in fixed broadband connections has helped to overcome the early problems with sound quality and high user expenses typical for telephone modem connections, but web radio and other streaming audio content on the Internet still have some disadvantages when compared to broadcasting. The main issue is economic: unlike in broadcasting, adding listeners over the Internet means more bandwidth expenses for the programme operator. This makes large-scale web radio operations expensive. The running costs of web radio services have been further increased by music royalties – and in some cases disputes over copyright issues force the stations to shut their web radio services or to restrict the use to certain geographic areas (Lax and Ala-Fossi 2008).

Another significant disadvantage of web radio has been its weak mobility and portability. Even at its best, WLAN (Wireless Local Area Network) coverage is very limited. However, there are an increasing number of solutions to make web radio more mobile. Dedicated wireless broadband networks are already available in some European countries,[7] as well as mobile broadband plans using mobile phone systems like GSM-based GPRS (General Packet Radio Service) and 3G cellular phone technologies like UMTS (Universal Mobile Telecommunications System), with fixed monthly rates becoming increasingly popular among consumers. Streaming audio from the Internet to a mobile handset or terminal using mobile phone networks has been possible for a while, but the data transfer with a regular 3G subscription has been quite expensive. According to one Finnish estimate the daily cost for web radio listening via a mobile phone could be about €2100 (Poropudas 2008) – and reception is always limited by the capacity of cellular network. However, mobile and wireless Internet radio reception in a moving car has been successfully tested. The first prototype for a dedicated Internet car radio using 3G networks was introduced in January 2009. Besides a mobile broadband plan, the users of this new device also must pay a separate fee for the radio itself (Radio Ink 2007; Moses 2009).

Another – and very different – way to overcome the problems with bandwidth cost and mobility in Internet audio delivery is podcasting. With suitable software you can subscribe audio files in advance to download from the Internet directly to your computer, portable player or mobile terminal. Podcasting has been described as a 'disruptive' technology (Berry 2006), but as an innovation it was architectural, bringing already existing elements together in a new way. This combination of digital audio files, publishing with RSS (Really Simple Syndication) feeds and portable digital music players was a rather marginal thing when first introduced in 2004. It has become a very successful new form of audio media among private citizens and broadcasters alike in a very short time. The role of podcast pioneer and former broadcast celebrity Adam Curry as the inventor of podcasting has been disputed, but he was certainly important for introducing it to the public (Newitz 2005; Berry 2006).

For an audio publisher, podcasting does not result in the same increasing bandwidth costs as streaming – and speech-oriented podcasts at least do not require any music royalty payments either. For the listener, podcast receiver software has made the use of Internet audio very easy: unlike previously, you no longer have to search and download every audio file separately to a computer and transfer the files onto the portable player. With podcast software you can make subscriptions so that whenever the pre-selected websites publish new audio files, the programme automatically downloads them and even places them in the player memory. Initially, in addition to a broadband connection, this also required that the portable audio player had to be connected to a computer. Now you can subscribe to podcasts directly from mobile wireless devices with a WLAN support. The consumption of podcasts can now be portable, mobile and independent of time and place – as well as free from problems of limited coverage and poor reception.

TV to go: and radio, too?

Although the Internet is by no means a broadcasting network, by the end of the 1990s it had became an important new delivery platform for broadcast radio (Lax and Ala-Fossi 2008), suggesting that the benefits of a converged network system were beginning to be understood in Europe also. By 1998, DAB was already considered to be outdated – some people even suggested that all digital broadcasting should be done using only DVB. However, these two systems were designed to be complementary and they could not substitute each other (Laven 1998; Wood 2001).[8] At the same time, NHK carried out extensive ISDB field trials in Tokyo (Yokohata 2007) and the developers of digital television in Europe noticed that the Japanese system could provide mobile television and multimedia reception for handheld terminals. Another EU-supported pan-European project [AC318 Motivate] was started in 1998. The goal was to compete with ISDB by developing a new European system for mobile television, later to be known as DVB-H (H for Handheld), adopted as an official standard in 2004 (Motivate 1998; Talmola 2005).

Nokia was one of the main initiators of and investors in DVB-H development (Torikka 2007). Besides bringing digital TV to mobile phones, DVB-H offers a relatively cheap and efficient high-data rate downstream channel to mobile terminals, which enables new kinds of video, audio and data subscription services for handheld devices. It is a modular innovation designed so that it could be deployed using existing DVB-T infrastructure, providing a converged IP-based platform between digital television and cellular phone networks like GSM and UMTS. It is possible to address the terminals individually by their IP numbers, provide users individual services, and also charge users according to the use of services or downloading content. Although DVB-H services can be implemented in DVB-T networks, it is not possible to receive DVB-T broadcasts with a handheld DVB-H receiver, or vice versa (Kornfeld and Reimers 2005; Sieber and Weck 2004; Lindqvist 2004).

The first commercial DVB-H services were launched in Italy in 2006, almost at the same time as ISDB mobile television (One-Seg) services were launched in Japan. Currently DVB-H services are available in twenty countries, some of which are also offering radio services (DVB-H 2008). Nokia has been actively promoting DVB-H as a system for digital radio (Ikonen 2008),[9] and interactive DVB-H radio services have also been tested in Finland (Kaasinen et.al. 2008). The EU has officially recommended DVB-H as the European standard for mobile television (EU 2008). The largest mobile TV market in Europe is Italy, where the three DVB-H operators have altogether about a one million paying subscribers. But in general, the success of DVB-H has been more modest than expected. In addition, when DVB-H was designed, DVB-T receivers consumed far too much power for mobile use – however Korean LG Electronics, for example, has introduced mobile phones with in-built DVB-T receivers for regular free-to-air digital broadcast television. Similarly, in Germany some operators have abandoned DVB-H for DVB-T in mobile use (O'Brien 2008).

In addition to a separate digital radio system, the US broadcasting industry had also in the late 1990s developed its own digital television standard called ATSC (Advanced Television Systems Committee), which was originally not designed or suitable for mobile use (ATSC 2006).[10] This was adopted by South Korea in 1997 who, having adopted NTSC in the analogue era, decided to stick with the US television standard in the digital age. When the Korean government wanted to launch mobile television services a couple of years later, neither ISDB nor DVB were a viable option. Instead of trying to develop ATSC-based mobile digital television, Korea decided in 2002 to enhance and update the earlier DAB multimedia concept by the Bosch group – and use those VHF frequencies that were originally reserved for DAB radio. In 2005, the new DMB standard for mobile TV was approved and first services were launched in Korea (Lee and Kwak 2005; Tunze 2005). So far, 8.2 million mobile phones with a DMB receiver have been sold in South Korea, making it the second largest mobile TV market in the world (O'Brien 2008).

Although DMB offers a technological multimedia update to DAB, it is still based on the traditional broadcasting paradigm: it gives you free access to all services and content. The DMB approach does not make it possible to identify individual users, which would allow direct charging of customers for content and services (Teng 2005; Engström 2004; Galbraith 2005). From a social point of view, this is a positive characteristic which results in freer communication – but from a business perspective, it can be seen as a weakness, preventing more efficient business models.[11] However, the cases of ISDB as well as DMB suggest that the access to free broadcast content might actually be the key to the success of mobile TV. According to recent study, about 330 million people will own a broadcast TV-enabled mobile handset in 2013, but less than 14% of them are willing to pay any extra for the service (Juniper Research 2008).

The second generation of DAB

By 2005, the original DAB system – and especially its audio coding – was widely considered to be inefficient and outdated. New systems like DVB-H, with more efficient audio coding, were able to deliver higher quality sound with lower bitrates or, equally, more services in the same amount of bandwidth. The audio coding issue was becoming more and more difficult because the relative success of DAB, especially in the UK, made it impossible to abandon the original audio coding, which would have made all existing DAB receivers obsolete – and on the other hand, new countries were no longer eager to adopt DAB because the relative inefficiency of the system. In 2005, Australia became the first country in the world to decide that DAB would be implemented only when it incorporated a more efficient audio coding system (Coonan 2005) such as MPEG-4 AAC, already supported by several other digital broadcasting standards like DRM, DMB, ISDB and DVB-H.

The political and commercial pressure to update the system was so high that in 2006 the World DAB Forum decided to adopt MPEG-4 HE-AAC v2 audio codec as an additional DAB standard, called DAB+ (Ross 2006). In addition, the organization which had been promoting DAB technology since 1995 also changed its name to the World DMB Forum, where DMB stands for a DAB-variant for mobile television and multimedia broadcasting. Although the official reason for changing the name was to avoid confusion and assumptions that the organization would represent only digital audio (World DMB 2006), it also clearly indicated the end of an era for the original DAB standard. Now there was hardly any reason why countries without an existing DAB provision should stick to the original version of DAB. Confirming this, France decided to adopt DMB and DRM as the preferred standards for digital radio (DRM 2007; World DMB 2007).

DAB+ is an incremental innovation; a more efficient version of the earlier system. While DAB is able to deliver five channels with CD-quality or nine channels with near FM-quality in a multiplex, DAB+ is able to deliver up to 28 near FM-quality channels in a single multiplex because similar sound quality can be reached with much lower bitrates. The French decision to select DMB instead of DAB+ was a very interesting move because DMB was not recommended for radio services by WorldDMB. It is less efficient than DAB+ and it is also missing many of the additional text- and data-related functionalities of DAB (Prosch 2007). However, the mobile multimedia capability of DAB and its variants were now arousing more interest than perhaps ever before. In the UK, a mobile TV service using so called DAB-IP was introduced in 2006 (and shut down a year later) (Meyer 2007), while some stations started to test DAB MOT, a system for delivering slideshows with audio to DAB multimedia receivers (Maitland 2007; Miles 2008).

No single standard, but at least a common receiver for a common market?

The genie had now been let out of the bottle. Suddenly, after 2005, DAB was not only competing with all the other digital systems but also with its own derivatives, which were obviously considered as substitutes to each other as well. In 2008, as noted in Chapter One, World DMB and EBU created new, Europe-wide digital radio receiver profiles. This was intended to prevent any further fragmentation of the digital radio market – and to support DAB as a multimedia platform – as well as the wider adoption of all DAB-related broadcasting systems at the same time. The so called *standard receiver* (profile 1) is meant for traditional type RDS-assisted audio reception with FM and AM, as well as DAB and DAB+. Both the *rich media receivers* (profile 2) and *multimedia receivers* (profile 3) should be capable of receiving all visually oriented additional DAB services – and the most sophisticated digital radio sets will be also able to receive DMB mobile TV! (World DMB 2008)

These profiles were a step towards a more pan-European approach, but they were also exactly what you might have expected the proponents of DAB-related broadcasting technologies and an alliance of European public broadcasters to find politically appropriate together. What was striking about these 'specifications for digital radio receivers for Europe' was the fact that they did not have any provision for DRM or the forthcoming DRM+ as accepted forms of digital radio broadcasting in Europe. In addition, why should a digital radio receiver be a dedicated piece of hardware with only limited connections (no WLAN, no USB) to the already existing universe of digital audio?[12] However, it seems that the receiver manufacturers have been much less suspicious towards also including non-broadcast Internet audio (and multimedia) connections into their latest models. After all, those are the only truly global standards for receiving digital radio services at the moment[13] – and perhaps also in the years to come.

Notes

1. An earlier version of this chapter was published in part in the *Journal of Radio and Audio Media* (Ala-Fossi et al 2008).
2. Fagerjord & Storsul (2007) have categorized the dominant interpretations of media convergence as convergence of networks; terminals; services; rhetorics; markets and regulatory regimes. They also think that convergence is a more useful concept to describe what we have already seen than what we now seeing now (complexity).
3. The first phase of Eureka 147 project took place in 1987–1991, and the second in 1992–1994. (Rissanen 1993)
4. According to a recent BBC trial, DRM on AM seems to share also serious interference problems at night with the AM version of IBOC (Plunkett 2009b; Stimson 2007).
5. WorldSpace had only 171,000 subscribers in 2008. (Clabaugh 2008)
6. HD Radio does not stand for High Definition, Hybrid Digital or anything else: as a matter of fact, iBiquity has admitted that "it is simply the branding language for this new technology" (http://www.hdradio.com/faq.php).

7. For example, in Finland @450 is a dedicated wireless broadband network, which is using Flash-ODFM – so it is not a mobile phone network.

8. According to a statement by the EBU Technical Committee, "the assumption that DVB-T can replace DAB, and that DAB is no longer needed, is not correct and should not be pursued by EBU members. Radio is a mobile and portable medium, and the best technology for it is DAB." (Kozamernik 1999) (cf. Wood 2001). Interestingly enough, as soon as the new DVB-T2 standard (designed for HDTV) was formally approved in 2009, it was also suggested to be a replacement for DAB+ (Gulbrandsen 2009).

9. In 2005, Nokia introduced Visual Radio, a hybrid system based on FM radio and synchronized visual GPRS data delivery via GSM network. The idea was that people using analogue radio through their Nokia handsets would be able to get text, pictures and interactive services related to audio content by paying the data transfer – or to buy additional data downloads. From 2005 until 2008, most of the new Nokia mobile phone models were equipped with Visual Radio functionality, while radio stations in almost 10 countries launched Visual Radio services (Elliot 2005; HP 2005; RCS 2007; Nokia 2006). However, the take up among consumers was very limited – and in November 2008, Nokia ceased development and removed all Visual Radio pages from the Internet. Another digital hybrid radio system for mobile phones introduced in 2005, iRadio by Motorola was tested, but never commercially released (http://broadband.motorola.com/iradio/).

10. ATSC did not even start development for mobile and handheld TV standard before 2007, and in practice the ATSC-M/H (mobile/handheld) system does not exist yet (ATSC 2008). First mobile TV services in the US were launched by Verizon Wireless in 2007 using the proprietary MediaFLO system developed and owned by Qualcomm, but so far they have less than 100,000 paying subscribers. AT&T launched a similar type of service in 2008 (O'Brien 2008). In addition, China has developed its own mobile TV standard, CMMB (China Multimedia Broadcasting) which is already operational in 150 Chinese cities (Jinglei 2009).

11. As a result, a special development project for marrying DMB-type multimedia broadcasting with IP technology called DXB (Digital Extended Broadcasting) was set up in 2005 (Twietmeyer & Dosch 2005). Later, another similar type of concept was introduced as DAB-IP (Meyer 2007).

12. DRM was considered during the development process of receiver profiles, but it was not included as a mandatory feature, because 'consumers would not be willing to pay more for their radios, in effect subsidising technologies they may be unable to use' (Howard 2009). However, the decision not to give even a recommendation for DRM support (or for WLAN and USB, either) is very hard to justify based on this argument.

13. There are at least 20 different codecs for Internet audio streaming, and usually receivers support only some of them. A new organization formed by radio manufacturers and broadcasters called Internet Media Device Alliance (IMDA) has just recently proposed the first global certification standard for Internet radios. It is estimated that certified devices would be able to receive about 90% of radio services currently available on the Internet. (Careless 2009).

References

Åberg, C. (2001), 'Diskreta ljud: om DAB som radio/Discreet sounds: About DAB as radio', *MedieKultur*, 33, pp. 44–61.

Ala-Fossi, M. and A.G Stavitsky (2003), 'Understanding IBOC: Digital Technology for Analog Economics', *Journal of Radio Studies*, 10:1, pp. 63–79.

Ala-Fossi, M., S. Lax, B. O'Neill, P. Jauert and H. Shaw (2008), 'The Future of Radio is Still Digital – But Which One? Expert Perspectives and Future Scenarios for Radio Media in 2015', *Journal of Radio & Audio Media*, 15:1, pp. 4–25.

Alcatel Space (2001), 'Contribution to the Internal Report of the technical Group of the Advanced Satellite Mobile System Task Force', *Alcatel Space*, August 2001, ftp://ftp.cordis.europa.eu/pub/ist/docs/ka4/asms_04_t16_0.doc. Accessed 25 May 2009.

Arango, T. (2008), 'Satellite Radio Still Reaches for the Payday', *The New York Times*, 26 December 2008, http://www.nytimes.com/2008/12/28/business/media/28radio.html?partner=rss. Accessed 25 May 2009.

ASTRA (2009), 'SES-ASTRA Channel guide', http://www.ses-astra.com/consumer/en/channel-guide/index.php. Accessed 25 May 2009.

ATSC (2006), 'ATSC Standard: Program and System Information Protocol for Terrestrial Broadcast and Cable (Revision C) With Amendment No. 1', Washington: Advanced Television Systems Committee, http://www.atsc.org/standards/a_65cr1_with_amend_1.pdf . Accessed 25 May 2009.

ATSC (2008), 'ATSC M/H Standard Developing At Rapid Pace: Advanced Television Systems Committee Receives Report from the Open Mobile Video Coalition (OMVC)', Press Release, 15 May 2008, http://www.atsc.org/communications/press/2008-05-15-idov-report-response.php. Accessed 25 May 2009.

BBC (2004a), 'Ceefax marks 30 years of service', *BBC News*, 22 September 2004, http://news.bbc.co.uk/1/hi/entertainment/tv_and_radio/3681174.stm. Accessed 25 May 2009.

BBC (2004b), 'DCMS Review of DAB Digital Radio – The BBC Submission', October 2004, http://www.bbc.co.uk/foi/docs/receiving_bbc_services/digital_services/DCMS_Review_of_DAB_Digital_Radio_-_DCMS_18Oct04.htm. Accessed 25 May 2009.

Beatty, P. (1998), 'Coping with Convergence: Social and Cultural Change in the Age of Digital Technology', Lecture to the University of Western Ontario (London), http://www.cbc.radio-canada.ca/speeches/19980320.shtml. Accessed 25 May 2009.

Berry, R. (2006), 'Will the iPod Kill the Radio Star? Profiling Podcasting as Radio', *Convergence: The International Journal of Research into New Media Technologies*, 12:2, pp. 143–162.

Careless, J. (2009), 'The Quest for Internet Radio Standards'. *Radio World*, 26 October 2009, http://www.rwonline.com/article/89328. Accessed 30 October 2009.

Clabaugh, J. (2008), 'WorldSpace bankruptcy plan gets court approval', *Washington Business Journal*, 22 October 2008, http://www.bizjournals.com/washington/stories/2008/10/20/daily62.html?ana=from_rss. Accessed 25 May 2009.

Commission of the European Communities (2008), 'Mobile TV across Europe: Commission endorses addition of DVB-H to EU List of Official Standards', Press release, 17 March 2008, Brussels: European Commission, http://europa.eu/rapid/pressReleasesAction.do?reference=IP/08/451. Accessed 25 May 2009.

Coonan, H. (2005), 'Speech by Australian Minister for Communications, Information Technology and the Arts about Digital Radio', *Commercial Radio Australia Conference*, 14 October 2005, Sydney, Australia, http://www.minister.dcita.gov.au/media/speeches/digital_radio_-_commercial_radio_australia_conference. Accessed October 26, 2005.

DRM (2005a), 'DRM Votes To Extend its System to 120 MHz: DRM and DAB Digital Radio Systems join forces to ensure digital radio solutions worldwide', Press release, Geneva and London, DRM Consortium and World DAB, 10 March 2005, http://www.worlddab.org/images/drmdabjointrelease-10-03-05.pdf. Accessed 25 May 2009.

DRM (2005b), 'DRM: Technical Aspects of The On-Air System', http://www.drm.org/system/technicalaspect.php. Accessed 7 September 2005.

DRM (2007), 'France Adopts DRM For Frequencies up to 30 MHz', Press release, 11 December 2007, Geneva: DRM Consortium, http://www.drm.org/uploads/media/press_release_147.pdf. Accessed 25 May 2009.

DRM (2008a), 'Positive Results From DRM+ Tests on FM', Press release, 29 May 2008, Geneva: DRM Consortium, http://www.drm.org/uploads/media/290508_Positive_Results_From_DRM__Tests_on_FM_E.pdf. Accessed 25 May 2009.

DRM (2008b), 'DRM+ Digital Radio Mondiale: A presentation', http://www.drm.org/fileadmin/media/downloads/drmplus_presentation_v1_5_.pdf . Accessed 21 October 2008.

DRM (2008c), 'BBC & DW in Europe on DRM Digital Radio', Press release, 10 December 2008, Geneva: DRM Consortium, http://www.drm.org/uploads/media/BBCDW_DRM_Channel_release_02.pdf. Accessed October 21, 2008.

DVB (2008a), 'DVB Timeline: History of DVB', http://www.dvb.org/about_dvb/history/dvb_timeline. Accessed 25 May 2009.

DVB (2008b), 'DVB Worldwide: Brazil', http://www.dvb.org/about_dvb/dvb_worldwide/brazil/index.xml. Accessed 25 May 2009.

DVB-H (2008), '[DVB-H] Services', http://www.dvb-h.org/services.htm. Accessed 25 May 2009.

Elliot, P. (2005), 'Mobiles get set for visual radio', *BBC News,* 21 January 2005, http://news.bbc.co.uk/2/hi/technology/4192351.stm. Accessed 25 May 2009.

Engström, K. (2004), 'Frequency Economy – New Convergence', *EBU Technical Review*, http://www.ebu.ch/departments/ technical/trev/trev_298-engstrom.pdf. Accessed 25 May 2009.

ESA (2007), 'Multimedia car radio of the future', Press release, 25 January 2007, Paris: European Space Agency, http://www.esa.int/esaCP/SEM9OBSMTWE_index_2.html. Accessed 25 May 2009.

Eureka (1986), 'Eureka 147 Project Form', http://www.eureka.be/inaction/AcShowProject.do?id=147. Accessed 25 May 2009.

Fagerjord, A. and T. Storsul (2007), 'Questioning Convergence', in T. Storsul and Dagny Stuedahl (eds.), *Ambivalence Towards Convergence,* Göteborg: Nordicom, pp. 19–32.

Fraunhofer IIS (2009), 'The Story of MP3', Fraunhofer IIS, http://www.iis.fraunhofer.de/EN/bf/amm/mp3history/mp3history01.jsp. Accessed 25 May 2009.

Frenzel, L. (2003), 'Satellite Radio gets Serious', *Electronic Design Online*, 18 August 2003, http://electronicdesign.com/Articles/Index.cfm?AD=1&ArticleID=5603. Accessed 25 May 2009.

Furukawa, H. (2005), 'Riding the Satellite TV Wave on Mobile Devices', *Japan Media Review*, 31 January 2005, http://www.japanmediareview.com/japan/stories/050131furukawa. Accessed 25 May 2009.

Gandy, C. (2003), 'DAB: an Introduction to the Eureka DAB System and a Guide to How it Works', BBC R&D White Paper, WHP 061, http://www.bbc.co.uk/rd/pubs/whp/whp061.shtml. Accessed 10 April 2009.

Galbraith, M. (2005), 'T-DMB takes on S-DMB in Korea', *Wireless Asia*, April 2005, http://www.telecomasia.net/telecomasia/article/articleDetail.jsp?id=158158. Accessed 25 May 2009.

GSM World (2008), 'History: A Brief history of GSM and the GSMA', http://www.gsmworld.com/about-us/history.htm. Accessed 25 May 2009.

Gulbrandsen, Per (2009), 'Kulturministern: Radio kan sändas i digital-TV-nätet / Minister of Culture: Radio can be delivered over the digital TV-network', Sveriges Radio, 29 September 2009, http://www.sr.se/sida/artikel.aspx?programid=1012&artikel=3132830. Accessed 16 October 2009.

Hallet, L. (2005), 'DRM Expands Into FM Sphere', *Radio World International Edition*, June 2005, pp. 1–3.

Henderson, R. and K. Clark (1990), 'Architectural Innovation: The Reconfiguration of Existing Product Technologies and The Failure of Established Firms', *Administrative Science Quarterly*, 35:1, pp. 9–30.

Hendy, D. (2000), 'A Political Economy of Radio in the Digital Age', *Journal of Radio Studies*, 7:1, pp. 213–233.

Henten, A. and R. Tadayoni (2008), 'The impact of Internet on media technology, platforms and innovation', in L. Küng, R. Picard, and R. Towse (eds.), *The Internet and the Mass Media*, London: Sage, pp. 45–64.

Howard, Q. (2009), 'One Digital Radio Across Europe', *EBU Technical Review*, http://www.ebu.ch/en/technical/trev/trev_2009-Q1_DAB-Rx.pdf. Accessed 30 May 2009.

HP (2005), 'Kiss FM aloitti Visual Radio -lähetykset tänään / Kiss FM launched Visual Radio broadcasts today', Press release, 4 March 2005, Helsinki: Hewlett-Packard Finland, http://h41131.www4.hp.com/fi/fi/press/Kiss_FM_aloitti_Visual_Radio_-lhetykset_tnn.html. Accessed 25 May 2009.

Ikeda, T. (2006), 'Transmission System for ISDB-TSB (Digital Terrestrial Sound Broadcasting)', *Proceedings of the IEEE*, 94:1, pp. 257–260.

Ikonen, A. (2008), 'DVB-H – Platform for digital radio', Presentation at *Las Vías para la Digitalización de la Radio en España*, 31 March 2008, Madrid, Spain, http://www.radiodigitaldab.com/descargas/4DVB-HradioMadrid31032008-SrIkonen.pps. Accessed 25 May 2009.

Immink, K. (1998), 'The CD Story', *The Journal of the AES*, 46:5, pp. 458–465.

Immonen, T. (1999), 'DAB (Digital Audio Broadcasting) – Digitaaliradio', http://www.tml.tkk.fi/Studies/Tik-110.300/1999/Essays/dab.html. Accessed 25 May 2009.

Jinglei, H. (2009), 'CMMB mobile TV signals reach 150 cities', Interfax TMT China, 20 January 2009, http://tmt.interfaxchina.com/news/1233. Accessed 25 May 2009.

Juniper Research (2008), 'Consumer spending on mobile broadcast TV to reach $2.7 billion by 2013, but revenues to be hit by free to air services, according to Juniper Research', Press release, 7 October 2008, http://www.juniperresearch.com/shop/viewpressrelease.php?pr=110. Accessed 25 May 2009.

Kaasinen, E., T. Kivinen, M. Kulju, L. Lindroos, V. Oksman, J. Kronlund and M. Uronen (2008), 'FinPilot2 Final Report. User Acceptance of Mobile TV Services', Helsinki: Forum Virium Helsinki, http://www.finnishmobiletv.com/press/FinPilot2_Final_Report_20080529.pdf.

Kato, H. (1998), 'Japan', in Anthony Smith (ed.), *Television: An International History; Second Edition*, Oxford: Oxford University Press, pp. 169–181.

Kim, K. (2003), 'DMB in South Korea', Presentation at 9th Meeting of the World DAB Forum General Assembly, 9–10 October 2003, http://www.worlddab.org/images/WorldDAB-398.pdf. Accessed July 18 2005.

Kornfeld, M. and U. Reimers (2005), 'DVB-H- the emerging standard for mobile data communication', *EBU Technical Review*, http://www.ebu.ch/trev_301-dvb-h.pdf. Accessed 25 May 2009.

Kozamernik, F. (1995), 'Digital Audio Broadcasting – radio now and for the future', *EBU Technical Review*, http://www.ebu.ch/en/technical/trev/trev_265-kozamernik.pdf. Accessed 25 May 2009.

Kozamernik, F. (1999), 'Digital Audio Broadcasting – coming out of the tunnel', *EBU Technical Review*, http://ebu.net/fr/technical/trev/trev_279-kozamernik.pdf. Accessed 25 May 2009.

Kozamernik, F. (2004), 'DAB- from Digital Radio towards Mobile Multimedia', *EBU Technical Review*, http://www.ebu.ch/trev_297-kozamernik.pdf. Accessed 25 May 2009.

Kuusela, S. (1996), 'DAB mullistaa radiokulttuurin/DAB will overturn radio culture', 22/96, 12 December 1996.

Kuusisto, P. (ed) (1998), *Kansallinen multimediaohjelma 1995–1997: Loppuraportti /National Multimedia Programme 1995–1997: Final report*, Technology programme report, 5/98, Helsinki: Finnish Funding Agency for Technology and Innovation.

Krueger, D. (2009), 'Letter from the CEO', Madrid: Ondas Media, http://www.ondasmedia.com/ceo.htm. Accessed 19 October 2009.

Laven, P. (1998), 'DAB- is it already out of date?' *EBU Technical Review,* http://www.ebu.ch/ departments/ technical/trev/trev_278-laven.pdf. Accessed 25 May 2009.

Lax, S. and M. Ala-Fossi (2008), 'The impact of the Internet on business models in the media industries – a sector-by-sector analysis: Radio', in L. Küng, R. Picard and R. Towse (eds.), *The Internet and the Mass Media,* London: Sage, pp. 159–163.

Lax, S. (2003), 'The Prospects for Digital Radio. Policy and technology for a new broadcasting System', *Information, Communication & Society,* 6:3, pp. 326–349.

Layer, D. (2001), 'Digital Radio takes to the Road', *IEEE Spectrum,* July 2001, pp. 40–46.

Lee, S. and D. Kwak (2005), 'TV in Your Cell Phone: The introduction of Digital Multimedia Broadcasting (DMB) in Korea', Paper presented at the *Telecommunications Policy Research Conference,* Arlington, VA., http://web.si.umich.edu/tprc/papers/2005/449/TPRC%202005_Final_ DMB%20in%20Korea.pdf. Accessed 25 May 2009.

Lindqvist, M. (2004), 'DVB-H & mobiilit joukkoviestinpalvelut/DVB-H & Mobile Broadcasting Services', Presentation at the *FICORA Frequency Seminar,* 22 April 2004, http://www.ficora.fi/ suomi/document/ts04_MaLi.pdf. Accessed 25 May 2009.

Mackay, H. and G. Gillespie (1992), 'Extending the Social Shaping of Technology Approach: Ideology and Appropriation', *Social Studies of Science,* 22:4, pp. 685–716.

MacKenzie, D. and J. Wacjman (eds.) (1999), *The Social Shaping of Technology – Second Edition,* Maidenhead and Philadelphia: Open University Press.

Maitland, A. (2007), 'DAB radio soon to broadcast images', *Pocket-lint.co.uk,* 23 January 2007, http:// www.pocket-lint.co.uk/news/news.phtml/6340/7364/dab-radio-soon-broadcast-images.phtml. Accessed 25 May 2009.

Mäkinen, S. and T. Nokelainen (2003), *Teknologioiden analysointi / Assessing Technologies,* Tampere, Finland: Department of Industrial Engineering and Management, Tampere University of Technology.

Mäntylä, J. (2000), 'Digiradio mikron sisällä / Digital radio inside PC', *Tietokone,* February 2000, http://www.tietokone.fi/lukusali/artikkelit/2000tk02/BOSCH.HTM. Accessed 25 May 2009.

Menduni, E. (2007), 'Four steps in innovative radio broadcasting: From QuickTime to podcasting', *The Radio Journal – International Studies in Broadcast and Audio Media,* 5:1, pp. 9–18.

Meyer, D. (2007), 'BT ditches mobile TV service', *ZDNet.co.uk,* 26 July 2007, http://news.zdnet.co.uk/ communications/0,1000000085,39288247,00.htm. Accessed 25 May 2009.

Miles, S. (2008), 'Pure Digital hint at DAB radio with video', *Pocket-lint.co.uk,* 21 August 2008, http:// www.pocket-lint.co.uk/news/news.phtml/17088/18112/pure-digital-dab-radio-video.phtml. Accessed 25 May 2009.

Miyazawa, S. (2004), 'The Trend of Digital Broadcasting in Japan', http://www.dibeg.org/PressR/ seminar_in_indonesia2004/presentation-1.pdf. Accessed 25 May 2009.

Mosco, V. (1996), *The Political Economy of Communication. Rethinking and Renewal,* London: Sage.

Moses, A. (2009), 'Internet car radio a world first', *The Age,* 7 January 2009, http://www.theage.com.au/ news/digital-life/home-entertainment/articles/Internet-car-radio-a-world-first/2009/ 01/07/1231004091554.html. Accessed 25 May 2009.

Motivate (1998), 'AC318 Motivate – Overall description and project website', http://cordis.europa.eu/ infowin/acts/rus/projects/ac318.htm. Accessed 25 May 2009.

Mykkänen, J. (1995), 'Yleisradiotoiminnan strategiaselvitys: Radio ja televisio 2010/Strategy Report on Broadcasting: Radio and Television 2010', Liikenneministeriön julkaisuja, 45/95, Helsinki, Finland: Ministry of Transport and Communications.

Nakahara, S. (2003), 'Overview: ISDB-T for sound broadcasting- Terrestrial Digital Radio in Japan', http://www.rthk.org.hk/about/digitalbroadcasting/DSBS/ABU_AIR_2003/ses2.pdf. Accessed 25 May 2009.

Newitz, A. (2005), 'Adam Curry Wants to Make You an iPod Radio Star', *Wired*, 13:3, http://www. wired.com/wired/archive/13.03/curry.html. Accessed 25 May 2009.

Nokia (2006), 'Visual Radio backgrounder', http://www.visualradio.com. Accessed September 20 2008. [Pages were removed from the internet in November 2008.]

O'Brien, K. (2008), 'Mobile TV gathers momentum', *International Herald Tribune*, 4 May 2008, http:// www.iht.com/bin/printfriendly.php?id=12545226. Accessed 25 May 2009.

Olon (2002), *EUREKA! Een oplossing voor digitale kleinschalige radio/EUREKA! A Solution for Small-scale Digital Radio – A report for the Directorate-General of Telecommunications and Post by the Olon – the Dutch Federation of Local Public Broadcasters*, Nijmegen, The Netherlands: Olon, http:// www.olon.nl/olon_public/olon3614.html. Accessed 25 May 2009.

PEJ (2009), 'Audio', In *The State of the News Media 2009: An Annual Report on American journalism*, Washington D.C: Project for Excellence in Journalism, http://www.stateofthemedia.org/2009/ printable_audio_audience.htm?media=10&cat=2. Accessed 19 October 2009.

Picard, R. (1989), *Media Economics: Concepts and Issues*, Newbury Park: Sage Publications.

Plunkett, J. (2009a), 'Digital radio listening falls', *Guardian.co.uk*, 29 January 2009, http://www. guardian.co.uk/media/2009/jan/29/digital-radio-listening-falls. Accessed 25 May 2009.

Plunkett, J. (2009b), 'Digital medium-wave test runs into trouble after dark', *Guardian.co.uk*, 22 May 2009, http://www.guardian.co.uk/media/2009/may/22/bbc-trial-digital-radio-mondiale-medium-wave. Accessed 30 May 2009.

Poropudas, T. (2008), 'Nettiradiopäivä maksaa 2100 euroa kännykkäverkossa/Web radio day in mobile phone network costs 2100 euros', *It-viikko*, 26 February 2008, http://www.itviikko.fi/pdf/20085820. Accessed 25 May 2009.

Priestman, C. (2002), *Web Radio: Radio Production for Internet Streaming*, London: Focal Press.

Prosch, M. (2007), *DAB+ The additional audio codec in DAB*, London: WorldDMB, http://www. worlddab.org/public_documents/dab_plus_brochure_200803.pdf. Accessed 25 May 2009.

Radio Ink (2007), 'Internet Radio Hits the Road', *Radio Ink*, 15 November 2007, http://www.radioink. com/HeadlineEntry.asp?hid=140073&pt=todaysnews. Accessed 25 May 2009.

RCS (2007), 'RCS and Nokia to Deliver Next-Generation "Visual Radio" Experience for Mobile Customers', Press release, 25 October 2007, New York: RCS, http://www.rcsworks.com/en/ company/newsitem.aspx?i=1011&c=2039655. Accessed 25 May 2009.

RDS (2008), 'What is RDS? A Brief Introduction to RDS (Radio Data System for VHF/FM broadcasting)', http://www.rds.org.uk/rds98/whatisrds.htm. Accessed 21 July 2008.

Rissanen, K. (1993), 'Digital Audio Broadcasting (DAB)', *Tuike*, May 1993, pp. 11–14.

Rogers, E. M. (2003), *Diffusion of Innovations; Fifth Edition*, New York & London: Free Press.

Ross, C. (2006), 'Eureka 147 Adds Option For Coding', *Radio World International Edition*, December, p. 1.

Rudin, R. (2006), 'The Development of DAB Digital Radio in the UK: The Battle for Control of a New Technology in an Old Medium', *Convergence: The International Journal of Research into New Media Technologies*, 12:2, 2006, pp. 163–178.

Schiffer, M. (1991), *The Portable Radio in American Life*, Tuscon, AZ. and London: University of Arizona Press.

Sieber, A. and C. Weck (2004), 'What's the difference between DVB-H and DAB – in the mobile environment?', *EBU Technical Review*, July 2004, http://www.ebu.ch/en/technical/trev/trev_299-weck.pdf. Accessed 25 May 2009.

Soramäki, M. and K. Okkonen (1999), *Taloudellinen integraatio ja EU:n audiovisuaalinen politiikka/ Economic Integration and Audiovisual Policy of the EUl*, Series D, 45, Tampere, Finland: Department of Journalism and Mass Communication, University of Tampere.

Suk, M. (2005), 'It's the DMB Age!', *Korea Economic Trends*, 26 February 2005, pp. 4–7.

Stimson, L. (2007), 'WYSL Claims Nighttime Interference From WBZ; May Be First Official AM IBOC Nighttime Complaint', *Radio World*, 31 October 2007, http://www.radioworld.com/article/8032. Accessed 30 May 2009.

Takahara, K. (2006), 'Business Model Still Tenuous: TV programs go mobile as One Seg services begin', *The Japan Times*, 1 April 2006, http://search.japantimes.co.jp/cgi-bin/nb20060401a2.html. Accessed 25 May 2009.

Talmola, P. (2005), 'DVB-H Standard Overview', http://www.abo.fi/~jbjorkqv/digitv/DVB_H_Standard_2.pdf. Accessed 25 May 2009.

Teng, R. (1998), 'Digital Multimedia Broadcasting in Korea', White paper In-Stat, http://www.in-stat.com/promos/05/WP_korea.asp. Accessed 25 May 2009.

Torikka, M. (2007), 'Insinööripalkinto meni mobiilitelevisiolle / The Finnish Engineering Award to Mobile TV', *Tekniikka ja Talous*, 24 May 2007, http://www.tekniikkatalous.fi/ict/article38703.ece. Accessed 25 May 2009.

Tunze, W. (2005), 'The DMB Story: Digital Multimedia Broadcasting went from Germany to Korea- and is now coming back to Germany', *Deutschland Magazine*, 30 September 2005, http://old.magazine-deutschland.de/magazin/OZ-IFA_5-05_ENG_E4.php?&lang=eng&lang=eng&lang=eng. Accessed 25 May 2009.

Twietmeyer, H. and C. Dosch (2005), 'DMB & DXB: Multimedia via DAB. Potential and Limits', Presentation at *Multiradio Multimedia Communications 2005: Broadcast meets Mobile*, http://mmc05.hhi.de/Downloads/2_Technologies_I/Dosch_DMB_DXB.pdf. Accessed 25 May 2009.

Valaskivi, K. (2007), *Mapping Media and Communication Research: Japan*, Department of Communication Research Reports, Communication Research Centre, April 2007, Helsinki: University of Helsinki.

Wood, D. (1995), 'Satellites, science and success. The DVB Story', *EBU Technical Review*, Winter 1995, www.ebu.ch/trev_266-wood.pdf. Accessed 25 May 2009.

Wood, D. (2001), 'Is DVB-T A Substitute For DAB?', *EBU-UER publication*, March 2001, pp. 1–3.

World DAB (2005), 'New wave of DAB legislation and developments worldwide', Press release, 14 April 2005, http://www.worlddab.org. Accessed 12 September 2005.

World DMB (2006), 'World DAB Forum changes its name', Press release, 30 October 2006, http://www.worlddab.org/upload/uploaddocs/WorldDMB_PR_301006.pdf. Accessed 25 May 2009.

World DMB (2007), *DAB · DAB+ · DMB Building on Success: Aiding the implementation and roll out of the Eureka 147 Family of Standards*, Country report brochure, London: World DMB, http://www.worlddab.org/rsc_brochure/lowres/4/worlddmb_brochure_20070814.pdf. Accessed 14 August 2007.

World DMB (2008), 'Europe-wide Digital Radio Just Got Easier', Press release, 12 September 2008, London and Geneva: World DMB and EBU, http://www.worlddab.org/public_documents/EBU_WorldDMB_Digital_Radio_Receiver_Profiles_Press_Release_12Sept08.pdf. Accessed 25 May 2009.

World DMB (2009), *Global Broadcasting Update. DAB/DAB+/DMB*, January 2009, London: World DMB, http://www.worlddab.org/rsc_brochure/lowres/5/rsc_brochure_lowres_20090114.pdf. Accessed 25 May 2009.

XM (2007), 'Alcatel Space delivers first payload for XM satellite radio', Press release, 17 April 2000, Washington: XM Radio, http://xmradio.mediaroom.com/index.php?s=press_releases&item=1074.

Yokohata, K. (2007), 'ISDB-T Transmission Technology: Single transmission for fixed, vehicular, and handheld receivers', Paper presented at *Summit on Information and Communication Technologies*, 27–28 September 2007, Tashkent, Uzbekistan, http://ru.ictp.uz/summit2007/yokohata.pdf. Accessed 25 May 2009.

Yoshino, T. (2000), 'Looking back at the Development of ISDB', http://www.nhk.or.jp/strl/publica/bt/en/tn0005-2.html. Accessed 25 May 2009.

Chapter 3

'A Vision for Radio': Engineering Solutions for Changing Audiences - from FM to DAB

Stephen Lax

F ollowing a lengthy gestation, digital radio continues to struggle to find a place as a mainstream medium in more than a small number of countries. At the end of 2009, some fourteen years after limited domestic services began in those countries with what might be regarded as established services (Denmark, the UK and, perhaps, Norway), terrestrial digital radio is operating according to the DAB standard developed under the EU Eureka programme, and was the first standard intended for domestic broadcasting to gain approval from the ITU. As seen elsewhere in this volume, DAB is both criticised and defended: criticised by some broadcasters and other commentators as being outdated by newer, emerging standards and also unsuited to today's broadcasting landscape, but defended by other broadcasters, particularly in those countries with existing services, as the only standard that has been proven technologically and in the market (other standards operating only as trials).

Frequently reported as an obvious or even inevitable technological development and ultimately a replacement for analogue radio, examination of the published technical and policy literature on the origins of DAB's emergence reveals a history that is certainly technological, but also economic and in some measure ideological. As a technological solution to a perceived problem, DAB would be considered by most to be successful, but it is by no means clear what the nature of the 'problem' actually is. DAB is of course not the first technological innovation in radio, and one of the most significant was the introduction of FM. Other innovations more often overlooked include stereo radio in the 1960s and the data system known as RDS introduced in the late 1980s. The introductions of these earlier technologies demonstrate a number of similarities to the introduction of digital radio and it is instructive therefore to consider the emergence of DAB in the more general context of technological innovation in radio broadcasting.

On a superficial level, digital radio does seem to be a straightforward, logical technological development. Throughout the literature detailing its technological beginnings (mainly in the engineering press), frequent reference is made to changing audience tastes and expectations. In particular, as discussed in earlier chapters, in explanations of the logic of digital radio, a common observation is the introduction and, by the end of the 1980s, the commercial success of the compact disc. The CD was a domestic digital audio format which had been anticipated to succeed existing analogue formats – vinyl disc and magnetic tape – based on better audio quality and superior resilience to wear and tear. With the widespread adoption of the CD interpreted as consumer acceptance of and support for these advantages, in comparison the audio quality failings of analogue radio were seen as problematic if radio was to compete successfully for the audience's ears. One illustration is given by Lau and Williams: 'With the advent of digital technology the general public now expects the same quality of reception on their radio as they achieve with home-based digital reproduction equipment such as their compact disc player' (1992: 5).

A second, recurrent theme in the early arguments for digital alternatives to analogue radio is the changing place of listening. In contrast with fixed location reception, mobile listening, that is, via a car radio (a standard feature of a new car's equipment by the 1980s) and on the newly popular personal radio and tape playing devices such as the Sony Walkman, revealed the apparent poor performance of analogue FM radio. In pointing out FM's deficiencies in comparison with the promise of DAB, many authors readily pointed out that FM was in fact never designed with mobile reception as one of its objectives when services began on the very high frequency (VHF) band in the 1950s, and so in some senses FM has actually operated in less than ideal conditions for almost half a century (for example, O'Leary 1993: 20).

A straightforward reading, therefore, of the early technical literature on DAB suggests a compelling logic for its development. Contrasted favourably with analogue radio, its 1995 launch did indeed bring forth numerous comparisons with the 'revolution' of FM (for example Goddard 1995; Maheu 1995). Just as that innovation came to be associated with a dramatic enhancement in audio quality and an expansion in the number of stations, DAB promised a similar step change. However, the introduction of FM was itself no straightforward matter of replacing its AM forebear, and neither have innovations in radio technology in the half century that followed been unproblematic. It is in the context of this history that the emergence of DAB should be studied.

Changing audience expectations: a longer history

Like DAB, FM was widely claimed to offer a significant improvement in sound quality in comparison with the then universal AM modulation system. By the early 1940s, its technical superiority was established. One leading figure amongst US radio engineers, W R G Baker, suggested in 1943 that FM was 'so much better technically than the present regular broadcast system that it can't fail of acceptance' (cited in Slotten 1996: 686). Yet despite such advantages, for several decades it did indeed fail to be accepted as a replacement for AM. Those technical advantages which were so self-evident in the 1940s were not so clear-cut a decade later. Perplexingly, the emergence of portable radios and mobile listening after the Second World War created a technically more complex challenge for FM, yet at the same time made the case for its rapid introduction still more compelling. FM transmission was originally designed to be received by an aerial fixed at a point some ten metres above ground level, typically in the loft or on the roof of a house. After field tests in the early 1950s, horizontal polarisation (in which the electric field component of the electromagnetic wave oscillates in a horizontal rather than vertical plane) was the preferred transmission mode as it suffered less interference, and so the receiving aerial's elements were also fixed horizontally. These fixed aerials were also directional, in that they received a strong signal from one direction and rejected signals coming from elsewhere (including reflections from nearby buildings or hills, known as 'multipath signals') – hence interference from sources other than the direct wave from the intended transmitter would effectively be screened out. For those prepared to acquire and install the necessary receiving equipment (a new radio receiver and a correctly-aligned, roof-mounted aerial) the benefit of FM's superior audio quality would be clear enough.

However, after the Second World War, the development of new, miniature glass valves requiring less electrical power than their predecessors, followed a decade later by the incorporation of transistors into the design of all stages of receiver construction, allowed the radio receiver to become more portable. Miniaturisation permitted the manufacture of 'compact' receivers which formed the second set in many households, and might habitually be moved around the home following the listener (Hill 1995: 40). When the transistor replaced all valves in receiver construction, radios could run off standard batteries, become smaller still and thus were truly portable in that they needed no mains power supply. This new locus of radio listening, in common with the increasingly popular car radio, meant that receiving aerials could no longer meet the ideal conditions for FM reception; that is fixed high on a roof and pointing directly to the nearest transmitter. Instead, they were now much closer to the ground, they were necessarily omni-directional as reception was required irrespective of direction of travel, and they were invariably vertically rather than horizontally oriented (for both practical purposes and to achieve this omni-directional performance). Consequently, those 'early adopters' who did purchase the new portable FM receivers frequently found them difficult to tune, generating numerous complaints to broadcasters (Briggs 1995: 840).

As a consequence, the plans drawn up in the 1940s for the implementation of FM now required modification. In fact, few early transistor-based portables or car radios actually had capability for FM reception – the feature was only available, if at all, at a significant extra cost. This scarcity presented a dilemma for broadcasters and their engineers: on the one hand, their new FM services could, after all, continue to be aimed as planned at those with fixed aerials and receivers who were less exposed to the difficulties of reception; on the other hand the popularity of new, AM-only portable radios slowed the adoption of FM receivers. For example, following a UK launch in 1955, the BBC rolled out the FM service relatively quickly: by the end of 1959 most transmitters had been upgraded and 96.4% of the population was within range of the signals; but even ten years later, when coverage was over 99%, the corporation noted that only one third of households had any form of FM receiver (Redmond 1969). A similarly slow rise in the popularity of FM continued in the US: although FM services had begun there on the VHF band some ten years earlier than in the UK, it was not until after 1979 that FM finally achieved a higher share of listening than AM (Slotten 1996: 687).

Yet the rapid adoption of portable, AM-only transistor radios (and AM-only car radios) also compromised the quality of reception on AM (already acknowledged to be inferior to FM). Earlier fixed AM receivers had also been designed to use external aerials with screened leads, which could eliminate a lot of the background noise inherent in AM transmission. With the emergence of the portable radio, the receiving aerial was relocated to the inside of the casing, both reducing the intensity of the received signal and increasing exposure to noise. As a result, in the mid-1950s many broadcasters significantly increased the power levels of their radio transmissions to boost the level of received signals. This, however, compounded the difficulties: because AM mostly uses medium frequency (MF) wavelengths ('medium wave') which carry much further at night time than during the day, after-dark listeners in one country were now experiencing high levels of interference from transmissions originating in other countries, and so AM listening at night became degraded and difficult. Once more, the case for using FM on VHF rather than MF/AM was reinforced as these very high frequency waves do not propagate far and thus do not cause this interference. In 1969, the BBC's director of engineering, James Redmond, believed the after-dark interference on AM to be so bad as to suggest it be used only for a reduced form of speech radio, and to move all music onto the higher-quality VHF/FM system, although even by that time he felt obliged to acknowledge that VHF/FM reception remained problematic for portable or in-car reception (Redmond 1969: 5–6).

Five years later, Redmond again expressed puzzlement at the slow adoption of FM, even for fixed reception in the home. Despite its superior sound quality, he noted that 'changeover has been slower than anticipated'. The reasons, he suspected, were not related to any deficiency in the technology. Instead, in addition to the unanticipated popularity of AM-only portables and car radios, the beginnings of FM came at a time when television also began to become a medium of mass consumption, and so there was less radio listening at night when the interference occurred. A second reason was the simulcasting of radio programmes on FM and on AM rather than offering new programming on the new service: listeners would only be able to hear on FM what their AM receivers already gave them. However, he suggested that the emergence of stereo on FM, and the imminent possibility of quadraphony, would encourage more and more to purchase an FM receiver (Redmond 1974: 6).

This history serves as an illustration of how an apparently self-evidently superior technology pursued as a solution to a problem of audio quality did not automatically find favour with listeners, who (other than the emerging band of 'enthusiasts' described below) were apparently prepared to put up with inferior sound and were less inclined to adopt FM while it offered little new programming or competition with television in the evening. Indeed, other changes in radio to some extent sent the plans for these new services awry: the emergence of portable receivers demonstrated some technical imperfections with the new system (although these were addressed as time went on by, for example, changing the polarisation of FM transmissions to suit vertically mounted aerials, and improving tuning circuits in receivers). A mismatch is revealed between the broadcasters' and engineers' beliefs as to what was important to listeners, and the preferences and priorities of the vast majority of those listeners themselves.

However, the assumption that the audience would seek more from their radio was perhaps understandable. After the first post-War Radio Exhibition at Olympia, London in 1947, a leading article in *Wireless World* celebrated the success of the radio industry in demonstrating (particularly during wartime itself) that radio had become 'something more than a means of distributing entertainment'. It noted the significant development in the design, miniaturisation and build quality of radio receivers, 'even some of the least expensive broadcast sets'. But a new tendency was reported: 'the provision of broadcast receivers for the discriminating user with requirements and tastes not catered for by standardized productions' (*Wireless World* 1947: 407). In the same issue it described the results of technical equipment tests 'of interest to the high fidelity enthusiast', and reported from Olympia the emergence of 'high-fidelity' equipment including loudspeakers and magnetic tape recorders (although the tape recorder intended for domestic uses was only of 'medium fidelity'). In other words, an interest was growing as early as 1947 in sound quality and domestic audio. In 1950, the long playing record was released, spinning at 33 ⅓ rpm instead of the 78 rpm disc it replaced, and this helped boost sales of the radiogram – record players integrated with radio receivers and loudspeakers – while reel-to-reel tape recorders also grew in popularity from this time. A more discerning audio enthusiast could be expected to appreciate the advantages of FM radio and go to the effort of installing suitable aerials and leads. By the end of the 1950s, the reproduction of stereo sound was exciting interest amongst these same enthusiasts; the title of a book by D. Gardiner, *Stereo and Hi-Fi as a Pastime*, sums up the nature of the new pursuit (cited in Hendy 2007: 197). It was at these audio enthusiasts that broadcasters perhaps aimed their new services, believing that radio needed to keep pace with such developments in audio fidelity if it were not to be displaced by these new technologies.

Even so, the number of enthusiasts only ever formed a small proportion of radio listening as a whole. Even *Wireless World* derided at some length the minority pursuit of audio excellence. In 1961, following another radio exhibition, it asked itself who was attracted to such events (*Wireless World* 1961: 237):

First, if only because the noises produced for them are difficult to ignore, are the 'hi-fi' enthusiasts. Their preoccupation is with sound for its own sake. The reproducing equipment must be extended to its limits, and if it wilts under the strain by as much as half a decibel or exhibits any signs of a hangover the weakness must be diagnosed and remedied at all costs. This is (one is tempted to say 'should be') a solitary pursuit. No two 'hi-fi' enthusiasts have ever been found to agree that the job has been properly done, though each may claim that his favourite method has been successful. With success comes satiety, and having exhausted the list of friends who can be enveigled [sic] into listening to snatches of larger than life test recordings the pastime begins to pall and the 'hi-fi' enthusiast moves on to tuning sports cars.

This group was contrasted with 'the most important group of all, the reasonable layman' who simply wants decent reproduction at a reasonable cost, and it was this far larger group that no doubt hesitated to replace perfectly adequate AM receivers with the more expensive

FM variety. Equally, they were perhaps less likely to embrace, or even notice, another new development in radio: the launch of stereo transmission in the 1960s.

The emergence of the LP and magnetic tape recording presented new possibilities for reproduction and manipulation of music, exemplified most clearly perhaps in the establishment in 1958 of the BBC Radiophonic Workshop, which experimented with tape recording to create new electronic sounds. Although early broadcasters' experiments with two channel radio had never intended to be extended to domestic broadcasts, by the end of the 1950s it was seen as logical to consider how stereo techniques emerging in LP records and domestic tape machines, and generating much interest amongst enthusiasts, could be incorporated into radio (not least because the playing of records comprised much of radio's programming). In 1959, the Institution of Electrical Engineers held a convention on stereophonic sound recording, reproduction and its implications for broadcasting, while in that same year the Study Group X of the *Comité consultatif international pour la radio* (CCIR, which makes recommendations on standards to the ITU) met in Los Angeles to consider the matter. Despite the somewhat subjective nature of stereo 'imaging' and its relationship to realism, coupled with technical difficulties posed by the need to preserve a suitable mono transmission alongside any stereo signal, within a few years suitable coding techniques had been developed and regular stereo broadcasts began in the US in June 1961. Within sixteen months, some 170 US stations were broadcasting in stereo (Utter 1963: 147). Despite some difficulties in reaching agreement on an international standard for stereo FM, the EBU agreed on adoption of the US system and stereo transmission began in a number of other countries in the following few years (Prose Walker 1963). In the UK, the BBC began a regular stereo service in 1966 and within five years most Radio 3 programmes were in stereo, and most on Radios 2 and 4 by the following year (Pawley 1972: 434).

Unlike the introduction of FM, which offered notable audio improvement to the listener, there is little evidence to suggest that the industry expected the radio audience to embrace stereo radio with any great haste. Certainly, there is no sense of bewilderment, or even exasperation, amongst broadcasters at the slow adoption of stereo receivers. Chicago station WEFM's chief engineer described 'undue concern' among a few broadcasters at the lack of immediate interest, and suggested that just as television viewers in the early years only gradually adapted receiving aerials to rid their screens of 'ghost' images, so the adoption of stereo receivers would also come slowly with time (Utter 1963). Briggs suggests that given the early experimentation in the technique by both radio producers and engineers, there was a 'built-in BBC commitment' to its development rather than a response to any demand, a commitment merely spurred on by the launch of the stereo LP (Briggs 1995: 829). Certainly, the creative possibilities of stereo were explored by radio producers, particularly in drama production (although the first stereo broadcasts were centred, perhaps naturally following the stereo LP, on music). A recurrent explanation offered for the interest in stereo amongst radio producers was a perceived need to innovate in audio in order to differentiate radio from television, which by now was offering innovations of its own, notably colour, adding further to its realism and appeal. BBC Managing Director of radio, Ian Trethowan, argued this explicitly: 'stereo is as momentous a development to radio as colour is to television' (1970: 4). Subsequent experiments in quadraphonic sound, including trial broadcasts of concerts and plays in 1976 and 1977 (eventually abandoned), were driven as much by creative producers as technical considerations such as fidelity or realism, given the obvious and ultimately insurmountable practical obstacles to its implementation in the domestic arena (Hendy 2007: 197). Hence, while FM presented clear technical advantages to the listener and thus, for broadcasters, sufficient reason for its development, the introduction of stereo radio did not constitute a response to a self-evident problem. While undoubtedly welcomed by some listeners, the small minority of audio enthusiasts already dabbling in stereo, a more significant driver was a perceived need to innovate for its own sake, both in order not to 'fall behind' in comparison with other media technologies (colour television, magnetic tape recording), and also to enable new creative possibilities at a time of innovations in other media.

Changing locus of listening

If developments in FM and stereo revealed assumptions, albeit overestimated, about the audience's increasingly discerning ear and thus a demand for audio fidelity, a second presumption was that more listening would take place on the move, in a vehicle. The car, in particular, would become an important place for radio listening. Car ownership itself grew rapidly during the 1960s and 1970s, and during this time a car radio increasingly became a standard fitment.

At the same time, the compact audio cassette tape was introduced as a domestic recordable platform – a popular, simple alternative to the enthusiast's reel-to-reel machine. It opened up to a new audience a realm of portable and customisable listening. The listener could now compile personal lists of songs recorded onto tape from either vinyl disc or from the radio, and play as and when desired. Combined portable cassette recorder and radio sets emphasised the close relationship between the two technologies, and the coming of the 'personal stereo' in the form of the Sony Walkman added mobility to the mix. Coupled with sophisticated electronic circuitry (the Dolby system) to reduce noise, audio quality was quite reasonable and in-car tape players soon appeared. In-car radio, until that time a monopoly audio source in vehicles, now had to compete with the audio quality and versatility of tape (fast forwarding and reversing to skip songs, for example). In particular, the 'youth market', which had grown up in the hi-fi age, was seen as driving (so to speak) a demand for more hi-fi audio performance in vehicles. Yet car radio technology remained beset by the difficulties which persisted with mobile FM reception, and many car radios were still AM-only, a situation many regarded as unsustainable. After all, went the argument, faced now with a choice, which source were drivers more likely to choose for their listening pleasure – the poor quality of AM radio or the superior fidelity of Dolby-enhanced cassette tape? Thus, radio needed 'updating' and measures were developed to improve mobile FM radio reception, reducing interference from other frequencies and reflections. These included the use of electronic circuitry far more sophisticated than that used in home-based receivers: the emergence of integrated circuits in the 1970s made it possible to design radio receivers with electronic displays (initially these were light emitting diode, also seen at the time in futuristic technologies such as calculators and digital wristwatches; later, more efficient fluorescent and liquid crystal displays were used); stations could now be tuned by automated electronic scanning of frequency bands rather than the traditional manual rotation of a knob to move a pointer across the scale; and favourite stations could be saved using memory buttons to permit easy recall; phase locked loop circuits, again based on advanced electronics, helped keep a station tuned once selected; other circuits automatically switched the radio from stereo to mono FM reception as the signal weakened, eliminating the additional noise found on stereo signals (Cherry 1980). Elements of such systems eventually found their way into portable receivers for the home, but were developed first and foremost for use in car radios, where the problems of mobile reception were most apparent and where the initial higher cost was more readily borne by the driver.

A further disadvantage of radio listening on the move was tackled with the introduction of the Radio Data System (RDS), beginning initially in the late 1970s but only fully developed by the end of the 1980s. Prior to RDS, tuning to a particular radio station meant remembering (or guessing) the station's frequency, or tuning randomly and waiting for a station announcement; and travelling in a vehicle meant the listener was soon out of range of that frequency and having to retune (the coverage areas of FM transmission were much smaller than AM, so the problem was exacerbated in this transmission mode). RDS meant the station's name would be displayed on the car radio rather than its frequency and, moreover, the searching for that station's frequency would be automated on leaving one transmission area and entering another. RDS was developed by European broadcasters under the auspices of the European Broadcasting Union (EBU) and intended to be a pan-European service, but one that might also be adopted in other continents (Beale and Kopitz 1993). With new features such as searching by programme type (rather than programme name or station name), the vision was of a car driver being able to drive the length and breadth of Europe and, at the press of a button or two, select a news, jazz, or classical music service or whatever was desired. The development of a traffic news service linked to RDS confirmed the assumption of the car-borne, mobile listener as the key radio audience towards the end of the twentieth century. Once more, a new use for radio was found for an audience whose expectations were assumed to have become more sophisticated.

By the end of the 1980s, it was assumed that the evolution of the audience's preferences had reached a zenith: 'The public now expect to be able to receive a stereophonic sound service in vehicles on the move and on portable receivers, in addition to the fixed home installation for which the services were planned' (Pommier and Ratliff 1988: 349). There appeared to be two underlying assumptions about the future of broadcast radio: firstly, the changing demands of the audience which was becoming used to higher audio fidelity and thus perceived to be ready to abandon radio should it fall short of those demands; and, secondly, the growing significance of mobile reception, principally the car, as the locus of radio listening and thus a place where additional functionality might readily add to the utility of broadcast radio. DAB continued the logic of these assumptions, designed at the outset for high quality audio in all conditions, including mobile reception, together with a high degree of multimedia capability.

DAB: continuing a trend

The development of DAB under the Eureka programme was not the first foray into digital radio. The underlying principle of digital technologies, pulse code modulation, had been understood since the 1930s, and realised in telecommunications from the 1960s. The BBC had been using the technique to relay its radio feed between transmitters since the 1970s. During the same period it developed a digital audio system, NICAM, to broadcast stereo sound to television, but trials of NICAM for domestic radio services revealed familiar problems: good quality reception was available only with fixed, rooftop aerials, whereas reception on mobile receivers on the ground was all but impossible due to multipath reception. Digital radio services had also been developed for delivery by satellite in the 1980s. These were designed for reception on fixed satellite dishes and expected to attract the same market as the growing number of radio stations on analogue satellite television services. However, subscription radio was never popular and early versions, such as DSR, ended when the satellite relaying the system reached its end of life. Other services developed later in the US by Sirius and XM and by World Space, used proprietary standards which were developed around the same time as DAB, but, again, these are intended largely for fixed reception (the newly merged Sirius XM services being relayed to mobile and fixed receivers in the US by a network of over a thousand terrestrial transmitters – in effect they are terrestrial services).

The advantages of digital coding of analogue signals were therefore understood and ready to be incorporated into DAB. Again, early published articles describing the emergence of DAB identify the in-car listener as the most important focus of technical efforts. Certainly, many accounts describe FM as perfectly adequate for fixed reception, implicitly recognising the success of its development in the 1960s and 1970s. Thus, Price explains that in such circumstances multipath reception problems can be overcome: 'with care the FM system can then deliver a performance approaching that of the CD' (Price 1992: 131). For all, however, the failings of FM for mobile reception are simply too profound to leave unaddressed, yet the refinement of analogue technology, such as FM, has gone as far as it can: 'although numerous and extensive technical improvements have been made to these analogue systems over the years … significant further refinements are probably no longer possible' (O'Leary 1993: 19). Some suggest that an additional cause of poor FM reception was, ironically, its success in generating excessive demand for spectrum, so that the VHF band has become too congested, leading to increased interference. Pommier and Ratliff note the paradox of increasing demand for services of apparently diminishing quality: 'thus the FM broadcasting services, which once delivered a sound quality into the home which was second-to-none, are under threat of erosion of the quality deliverable' (1988: 349. See also O'Leary 1993, Shelswell 1995). With the expectation that a new digital radio service would be allocated new spectrum, a further incentive to digitalization was the expansion in capacity for radio stations.

However, the emphasis on high quality mobile reception was foremost. Hoeg et al.'s introductory comments for DAB industry professionals make this explicit in describing the motivation of the Eureka 147 project: 'Perfect mobile reception was the overall aim', even in cars travelling 'at high speeds' (Hoeg et al. 2001: 5). According to Gandy, it was this potential feature that was the decisive factor for the BBC's adoption of DAB as the future for its radio broadcasting (Gandy 2003: 2). As O'Neill observes in this volume, numerous early commentators referred to being able to drive across whole countries, or even the whole of Europe, listening to a single station without having to re-tune the radio (Fox 1991: 29; Tuttlebee and Hawkins 1998: 268; Hoeg et al. 2001: 3). The market for in-car digital radios was expected to be one of the first to develop, influencing the BBC's decision to roll out DAB services initially along motorway corridors (Gleave 1997: 240). The introduction of RDS, developed just a year or two ahead of DAB, had of course aimed to achieve the same effect in the analogue FM system in automating tuning and station selection as large distances are covered – one account of early market research on DAB's potential, suggesting that ease of tuning was a notable benefit to listeners, acknowledged that this was 'also true of RDS' (Müller-Römer 1997: 19). Even so, it was the phrases 'CD-quality sound' and 'digital audio' that continually distinguished digital from analogue radio, even though the same market research had implied that, for the listener, this might be of secondary importance to ease of tuning. The widespread adoption of CD players meant that for many, the CD was now the 'yardstick' by which audio quality would be measured (for example Pommier and Ratliff 1988; O'Leary 1993). Any source of lower-fidelity sound would now be regarded as somehow deficient and, notwithstanding Price's favourable comparison above, this was generally taken to include analogue FM (for example Lau and Williams 1992; Price 1992). As we shall see, there is little evidence to support this presumption but, nevertheless, early promotion of DAB by the industry certainly used the phrase CD-quality sound, and placed this and related phrases at the top of the list of DAB's advantages (Lax 2003: 335–6; O'Neill, this volume). Again, as with the introduction of FM, it was the introduction of a new technology in another arena (recorded sound) that was seen as impelling radio towards new developments. Other digital radio features, such as some multimedia capability, were generally noted but understated in any promotion, in part because it remained unclear quite how those facilities would be used. Even so, such features were still seen by some as a potential response by radio to the audience's changing expectations of its media technologies; for example, Kopitz and Marks refer to the 'higher expectations from our audiences' as they describe how DAB can offer a more sophisticated form of traffic information flow than analogue's RDS traffic announcement system (1999: 4).

Thus, from the early technical literature during the introduction of DAB, we can observe a tendency to continue themes established in the introduction and subsequent development of VHF and FM: a sense that the listener's tastes and expectations were changing, with exposure to other media threatening a reduction of interest in radio; and a continuing migration to the car as the most important location of listening. However, just as the slow pace of adoption of FM confounded broadcasters, for whom its advantages were self-evident, DAB too has failed to gain an enthusiastic embrace from the audience. If FM's superior sound was not the audience's main priority in the 1960s, when there was more interest in radio's new portability, the fact that its technical superiority was also not always evident (as noted, it still suffered from problems with portable receivers) simply compounded the difficulties in securing its acceptance. DAB too has its problems with claims for technical superiority: it is now widely acknowledged that the audio quality of DAB is no better than, or even inferior to, analogue FM when pushed to the limit (for example Spikofski 2003; Holm 2007); even for those who do not demand hi-fi quality, the configuration of DAB can mean some stations which broadcast in stereo on FM are transmitted only in mono on DAB, although this is due to the way DAB has been implemented rather than an inherent technical shortcoming. Similarly, the assumption that radio developments should place a greater emphasis on serving the car driver perhaps overlooked the fact that, typically, for every hour of radio listened to in a car four hours were listened to elsewhere (RAJAR 1999). The presumption of the importance of in-car reception in the DAB project has also not been borne out given that the number of DAB receivers in vehicles is very low (around 1.5% of the numbers bought for the home in the UK for example) with no car manufacturers fitting such sets as standard, citing problems with robustness of mobile reception (DRWG 2008). Again, while it can be argued that a different implementation of DAB (such as additional or higher power transmitters) would solve this problem, nevertheless for the everyday listening experience, like FM, the promise of technological supremacy for mobile reception has not been fully realised.

Engineering solutions, economic solutions

The accounts above, of the introduction and subsequent development of FM and of the emergence of DAB, are drawn mainly from the specialist electronics literature, and thus reflect strongly the perspectives of the engineers and technical writers, a perspective that informs most popular accounts of the development of radio. Certainly, many commentators, both within the radio industry and external observers, see progression to digital technology as an obvious and essential next step. The exhortation, 'radio *must* go digital', has been expressed repeatedly over the years: for example, by the Director General of Audio Visual Policy at the EU's Media programme, Spyros Papas, in 1998; by BBC Director of Radio, Jenny Abramsky in 2003; and more recently, in 2009, by French National Assembly member Patrice Martin-Lalande and, less surprisingly, by Quentin Howard of World DMB. For these commentators, the logic of this transition is self-evident and so needs little explanation, technical or otherwise, and none is offered – put simply, radio cannot remain an analogue technology when all other consumer technologies are digital. Yet however compelling the logic might be from a technical point of view, the development of both FM and of DAB have failed to follow it: both have emerged only slowly and, in the case of DAB, its future remains uncertain.

If the development of DAB is examined as more than a technical project, the reasons for its difficulties become more apparent. It can be seen that support for DAB was driven equally by economic, and to some extent, ideological considerations as by technical motivations. As we have seen elsewhere in this volume, at the outset, in seeking support from the Eureka programme, an economic case for developing DAB was at the heart of the project. The quest for approval of DAB as a standard proceeded quickly: joining together with the European Broadcasting Union, DAB was first reported to CCIR in 1987 in preparation for submission for standardisation, with an official presentation the following year at the World Administrative Radio Conference in Geneva (Rau et al. 1990). Despite the setback of the US rejecting the Eureka system (following an initial favourable response), the EBU and World DAB Forum's efforts to gain approval for DAB and its component technologies during the first years of the 1990s resulted successfully in an ITU recommendation in 1994 (Hoeg et al. 2001: 8). Finally, following the allocation of frequencies across Europe for terrestrial transmission in 1995, DAB was ready to begin. Aware that competitor systems were being developed in Japan and now in the US, broadcasters (through the EBU) and manufacturers urged the EU to support DAB in the same way that the GSM mobile phone system had been supported.

As part of those efforts, DAB was celebrated as perhaps the first convergent 'information age' technology. As a digital platform, DAB had always been intended to be able to transmit data alongside audio information, and even the very first receivers had small screens to display scrolling text. At a 1998 European Commission meeting, industry speakers argued that DAB was effectively a mobile multimedia data-casting platform, one which was ideally suited to the needs of the information society. It was claimed that DAB 'recognises the mobility of our society more than any other broadcasting or telecommunications system. DAB recognises the need by our society for integrating different services in one system', while another speaker described DAB as 'a mobile information highway'. Again, the assumption was that radio had to adopt digital technology to take its place in the information society (EC 1998a). Embracing convergence, some of the earliest receivers were actually computer modules aimed at PC users, where DAB's multimedia capabilities could be exploited to the full. Despite the efforts of broadcasters and manufacturers, however, radio did not become part of the EU strategy for the audiovisual industry – in its 40 page report, published six months after the DAB meeting, the High Level Group on Audiovisual Policy refers to television on 90 occasions and to radio just once (only then in the context of delivery over digital television networks) (EC 1998b). DAB was not seen as a case under threat from other competing systems (the US had no pretensions that IBOC might become a global standard) in the way that digital television might have been. Equally, DAB as an information age, mobile multimedia platform became eclipsed by the enthusiasm (within and beyond the EU) for 3G mobile telecommunications networks, as indicated in 2000 by the vast sums bid by the telecoms industry for 3G licences at the height (and immediately prior to the end) of the dot-com boom. As O'Neill and Shaw point out (Chapter One), the EU support sought by the radio industry was never forthcoming and broadcasters were now forced to promote DAB in the open market. The growth of digital radio now depended much more on policies adopted in individual countries, and those policies (and thus the success of DAB) varied considerably (Lax et al. 2008).

In a similar way, the commercial success of the CD, cited frequently as a motivation for DAB's development, can be explained as due more to economic than technological factors, thus undermining the belief professed in the technical literature that the CD had cultivated (or simply satisfied) a more discerning listener demanding higher audio fidelity. After all, by the time of the CD's introduction, the audio cassette tape had become widely adopted. Never a 'high end' audio performer, in most circumstances the cassette offered inferior sound quality to its vinyl LP predecessor. Certainly the small speakers in portable cassette players and the still tinier headphones in the Sony Walkman and its emulators did not help. Instead, as noted earlier, the cassette offered two unique attributes contributing to its success: its portability and its recordability. Neither of these virtues was available to the CD when it emerged, and an economic reading of the CD's history suggests that this was an important deficiency. Winston argues that the CD's success is a 'curious history', the result of 'a completely effective manipulation of the market by a few international communication conglomerates' (1998: 135–6). For the music industry, the CD was a welcome shot in the arm at a time when sales of LPs had been falling for some years. Sales of pre-recorded cassettes partly accounted for this decline, but the industry suspected 'piracy' in the form of recording from LP onto cassette (though there is little evidence for this). As a non-recordable format, the CD was therefore favoured by the industry but shunned by consumers as offering little that was not already available with the LP, and consumer wariness was compounded by questions over the durability of the new discs (Horstmann and MacDonald 2003: 320). Early sales were slow: in 1984, a year after its widely publicised launch, just 800,000 CDs were sold in the US in comparison with 200m LPs. By selectively releasing music only in CD format and withdrawing LPs, music distributors ensured consumers were faced with little choice but to adopt the new format, and by 1995 the transition was complete – just 2.2m LPs pressed versus 700m CDs sold. Hence, the growing adoption of the CD player noted by DAB's proponents might imply little about sound quality but more about availability of recorded music.

A further difficulty for DAB was the changing landscape of radio in many countries during the period of its development, such that by the time of its public launch in 1995 it was seen by some as reflecting a view of the radio industry that was out of date. Developed largely under the auspices of the EBU from the 1980s, the configuration of the DAB system, based on ensembles of five or more stations grouped together in nationally transmitted, single frequency networks, was highly appropriate to a radio landscape in which public broadcasting networks were predominant. The BBC, for example, already transmitted five national analogue networks which accounted for the vast majority of radio listening in the 1980s and its regional and local stations were tiny in comparison. Once more, the image of a car driver covering vast distances while remaining tuned to a single station, as envisaged with both DAB and RDS (developed during a similar period), reflected this view of large area, networked coverage. Yet, in many countries, from the 1990s smaller commercial and community stations began to grow in number and equally succeeded in winning significant audience share. These stations, often independently-owned with small geographical coverage, would struggle to fit into a DAB system configured to favour larger coverage areas, and so the commercial and community radio sector has generally shown a lack of enthusiasm for DAB, adding to the difficulty in its adoption and eventual replacement of analogue radio (Lax et al. 2008). Just as by the time FM was launched, other changes in radio had made its introduction more complex, so too we can observe similar, non-technological reasons for the problems in introducing DAB.

These 'micro-histories' indicate the presence of a number of drivers of technological change in radio, a range of factors evident in innovations spanning half a century. Raymond Williams described the tension between technological and cultural explanations of television's development more than three decades ago. He argued then for including the notion of 'intention' in an explanation of technological change. That is, technology is 'being looked for and developed with certain purposes and practices already in mind' (Williams 1990: 14): with social needs so defined, technological solutions are sought. This is not to say that such intentions are always realised, implying some kind of social rather than technological determinism, and there are plentiful histories which demonstrate the complex trajectories taken by technological innovations. In the case of digital radio, it is possible to identify a number of intentions behind its development, from an imagined need to compete with other emerging technologies to a macro-economic need to aid a key industry. In contrast to the history of radio technology frequently presented as a straightforward series of technical challenges faced and solutions proffered, we find instead that apparently compelling innovations follow a complex path in which cultural practices and economic interests must be taken into account. As World DMB president Quentin Howard acknowledges, perhaps belatedly, the economic success of DAB will depend far more on attracting a mass audience than appealing to 'a small minority of audiophiles' (Howard 2009).

References

Beale, T and D. Kopitz (1993), 'RDS in Europe, RBDS in the USA – what are the differences and how can receivers cope with both systems?', *EBU Technical Review* 255, pp. 5–11.

Briggs, A (1995), *The History of Broadcasting in the United Kingdom Vol. 5: Competition,* Oxford: OUP.

Cherry, J. R. (1980), 'Electronic radios and hi-fi sounds', *Proceedings of 30th IEEE Vehicular Technology Conference,* 15–17 September, pp. 466–9.

DRWG (2008), *Digital Radio Working Group: Interim Report,* London: DCMS.

EC (1998a), *Radio in the Digital Era,* Conference organised by the EC DG X, Audio Visual Policy, Brussels, 5 March 1998.

EC (1998b), 'High Level Group on Audiovisual Policy', *The Digital Age: European Audiovisual Policy,* Brussels: European Commission.

Eureka (1986), *Eureka 147 Project Form,* Brussels: EU.

Fox, B. (1991), 'Radio sans frontieres', *New Scientist,* 20 July.

Gandy, C. (2003), *DAB: an Introduction to the Eureka DAB System,* BBC White Paper 61, London: BBC.

Gleave, M. (1997), 'Digital radio takes off', *IEE Review,* 43:6, pp. 239–42.

Goddard, P. (1995), 'Digital radio rattles music biz: It's possible you'll never need or want another CD', *Toronto Star* (Arts section), 18 February, p. 10.

Hendy, D. (2007), *Life on Air: a History of Radio Four,* Oxford: OUP.

Hill, J. (1995), 'The receiver', *EBU Technical Review,* 263, pp. 33–44.

Hoeg, W., T. Lauterbach, E. Meier-Engelen and H. Schulze (2001), 'Introduction', in W. Hoeg and T. Lauterbach (eds.), *Digital Audio Broadcasting: Principles and Applications,* Chichester: Wiley.

Holm, S. (2007), 'Audio quality on the air in DAB digital radio in Norway', *Proceedings of the 31st Audio Engineering Society conference,* 25–27 June, London.

Horstmann, I. and G. MacDonald (2003), 'Is advertising a signal of product quality? Evidence from the compact disc player market 1983–1992', *International Journal of Industrial Organization,* 21:3, pp. 317–45.

Howard, Q. (2009), 'One digital radio across Europe', *EBU Technical Review,* 317, pp. 1–11.

Kopitz, D. and B. Marks (1999), 'Traffic and travel information broadcasting – protocols for the 21st century', *EBU Technical Review,* 279, pp. 4–12.

Lau, A. and W. Williams (1992), 'Service planning for terrestrial Digital Audio Broadcasting', *EBU Technical Review,* 252, pp. 4–26.

Lax, S. (2003), 'The prospects for digital radio', *Information, Communication and Society,* 6:3, pp. 326–49.

Lax, S., M. Ala-Fossi, P. Jauert and H. Shaw (2008), 'DAB, the future of radio? The development of digital radio in four European countries', *Media, Culture and Society,* 30:1, pp. 151–66.

Lembke, J. (2003), 'Strategies, politics and high technology in Europe', *Comparative European Politics,*1:3, pp. 253–75.

Maheu, J. (1995), 'Broadcast on France Inter radio', *BBC Summary of World Broadcasts,* 30 November.

Müller-Römer, F. (1997), 'DAB progress report', *EBU Technical Review,* 274, pp. 12–22.

O'Leary, T. (1993), 'Terrestrial audio broadcasting in Europe', *EBU Technical Review,* 255, pp. 19–26.

O'Neill, B. (2009), 'DAB Eureka 147: a European vision for radio', *New Media and Society,* 11:2, pp. 261–78.

Pawley, E. (1972), *BBC Engineering 1922–1972,* London: BBC.

Pommier, D. and P.A Ratliff (1988), 'New prospects for high-quality digital sound broadcasting to mobile, portable and fixed radio receivers', *IBC 98 Conference Publication,* 23–27 September, pp. 349–53.

Price, H. (1992), 'CD by radio: digital audio broadcasting', *IEE Review,* 38:4, pp. 131–5.

Prose Walker, A. (1963), 'A resume of current activities of CCIR Study Group X of particular interest to audio engineers', *Journal of the Audio Engineering Society,* 11:4, pp. 2–6.

RAJAR (1999), *Radio listening in-car: RAJAR Insight 3,* London: RAJAR.

Rau, M., L. Claudy and S. Salek (1990), 'Terrestrial coverage considerations for digital audio broadcasting systems', *IEE Transactions on Broadcasting,* 36:4, pp. 275–83.

Redmond, J. (1969) 'Radio and television engineering: the next phase', *BBC Lunch-time Lectures, Seventh Series,* 19 March.

Redmond, J. (1974), 'Broadcasting: the developing technology', *BBC Lunch-time Lectures, Ninth Series,* 12 November.

Shelswell, P. (1995), 'The COFDM modulation system: the heart of digital audio broadcasting', *Electronics and Communication Engineering Journal,* 7:3, pp. 127–36.

Slotten, H. (1996), '"Rainbow in the sky": FM radio, technical superiority, and regulatory decision-making', *Technology and Culture,* 37:4, pp. 686–720.

Trethowan, I. (1970), 'Radio in the seventies', *BBC Lunch-time Lectures, Eighth Series,* 5 March 1970.

Tuttlebee, W. and D. Hawkins (1998), 'Consumer digital radio: from concept to reality', *Electronics and Communication Engineering Journal,* 10: 6, pp. 263–76.

Utter, R. (1963), 'Some practical aspects of FM stereophonic broadcasting', *Journal of the Audio Engineering Society,* 11:2, pp. 147–54.

Williams, R. (1990), *Television: Technology and Cultural Form; Second Edition,* London: Routledge.

Winston, B. (1998), *Media Technology and Society,* London: Routledge.

Wireless World (1947), 'Reflections on Olympia', November, p. 407.

Wireless World (1961), 'Recreation in sound', May, p. 237.

Chapter 4

'Sounding the Future': Digital Radio and CD-quality Audio

Brian O'Neill

C entral to the early effort to win acceptance for DAB in the early 1990s was an extensive process of promotion of the many claimed advantages of the new broadcasting technology. Digital radio broadcasting under the Eureka 147 DAB project offered many technical enhancements – more efficient use of the spectrum, improved transmission methods, and lower running costs – features that were attractive to industry professionals, broadcasting organisations, regulators and spectrum planners. But digital radio was also designed as a consumer proposition offering audiences a new and improved listening experience with ease of tuning, reliable reception, text and data services, interactive features, and significantly, 'CD-quality' audio. The promise of digital radio was to be 'the sound of future'. Notwithstanding ongoing debates about the actual audio performance of the DAB system, this chapter revisits some of the early claims for a radio listening experience of unsurpassed quality. An emphasis on DAB's audiophile credentials was, and continues to be, an important component of the marketing strategy for digital radio. This chapter contextualises DAB's promise to offer 'perfect sound', locating it within the broader historical context of digitalization and audio fidelity. The ambition to extend and improve radio was a central tenet of the founding vision for DAB, and a core element of this was a profound belief in the importance and perfectability of its sound. How such an emphasis has proved to be so fragile and whether this is out of step with listeners' expectations and experiences are questions addressed in the following.

The Sound of the Future

The development of Digital Audio Broadcasting (DAB), or Eureka 147, in the early 1990s was accompanied by a tremendous optimism bouyed by its technical achievements concerning the potential for innovative new dimensions to radio as a medium. A key feature of this was its promise of exceptional audio quality. The many promotional claims for its advanced and superior quality emphasised that DAB was the 'sound of the future', echoing previous historical breakthroughs in technology including the development of FM, the compact disc or, previously, the long playing vinyl record, which similarly promised a major advance in the quality of the audio signal, enhancing the listener's enjoyment and providing an experience of audio fidelity not previously available.

A Canadian government report in 1995, *The Sound of the Future,* for instance, proclaimed that:

Digital radio *is* the sound of the future. It will be the best sound on the airwaves before the end of this century because digital radio has the potential to deliver CD-quality audio, interference-free sound. (Task Force on Digital Radio 1995)

The trade press characterised DAB as the 'perfect sound machine' and likened its launch in the United Kingdom to the change to 625-line transmissions and the introduction of colour television in the 1960s, with a promise of hi-fi stereo-sound quality up to compact disc standard with the combined robustness of Long Wave (Fox 1994).

As noted by Ala-Fossi (this volume), the development of DAB Eureka technology should be seen within a broader process of digitalization that dominated technical broadcasting development in the late 1980s and early 1990s. As the intended replacement technology for AM and FM radio broadcasting (Hoeg and Lauterbach 2001), the system promised a host of innovations and benefits for both broadcasters and listeners. Using the newly developed digital techniques of audio encoding and compression, listeners could avail themselves of some of the best audio technology of the time, equivalent to that used in the Compact Disc format as well as in the digital stereo sound system for terrestrial and satellite television, and in other consumer audio products such as digital compact cassette and digital audio tape (DAT). While the immediate objective of DAB might have been to improve FM's susceptibility to interference, especially in mobile conditions (see Chapter Three), one of the consequences was unrivalled quality of audio in mobile reception conditions, making the car entertainment system the equivalent of a high-end home stereo system (Shelswell et al. 1991; Lau et al. 1992). In Chapter Three, Lax notes the frequent references to driving across Europe without the need to retune the radio. As one contemporary account has it:

Imagine driving the length of Britain, over the Channel and across Europe, listening all the time to the same radio station. The sound is in digital stereo, which gives it the same quality as that from a compact disc. There is no interference, and none of the fading and fluttering that normally blemish reception as you drive past tall buildings, over hills and down valleys. There is no need to keep retuning the radio because the chosen station remains on the same frequency throughout Europe – although, of course, you could retune to alternative national, international or local stations if you wanted to. (Fox 1991)

Market research conducted for the BBC in 1997 suggested that among the early adopters for digital radio would be hi-fi enthusiasts and serious music fans, as well owners of new gadgets such as surround sound TV, in-car CD equipment, and newer formats such as Digital Compact Cassette (DCC) (Tuttlebee and Hawkins 1998: 265). While new stations, dynamic text and visuals would act as purchase triggers for digital radio, improved sound quality, it was believed, would be a 'post-purchase' reward that would support DAB's long term acceptance and adoption. As a result, the audiophile credentials for DAB, and the appeal to the discerning standards of critical audio listeners, became a central part of the promotional discourse for the platform. A futuristic scenario from the technical press in the mid-1990s portrays the ideal listener, enjoying the benefits of DAB as a fully integrated digital entertainment system:

> Returning home from a business trip, Doug Digital turns on his car radio and enters code 15 for classical music. After the radio selects an appropriate strong-signal digital audio broadcasting station, Doug hums along, adding his voice to the compact-disk [sic] quality sound of the selection, which is free of any interference or signal fading despite the hilly terrain. He likes the music, but cannot put his finger on what it is, so he glances down at the radio's liquid-crystal display and reads the name of the selection and the performing artists. As he travels farther away from the station's transmitting facility, the radio switches to a stronger station airing the same classical programming, without his noticing the changeover.
>
> ... When Doug gets home, he and his wife have dinner and then decide to listen to a live concert of the New York Philharmonic orchestra. Doug requests the concert from the pay-per-listen digital audio radio service he subscribes to and the two settle back, listening to it in five-channel Dolby Surround on their stereo system. After the concert, Doug decides to add features to his digital audio radio system, including programming it with a "pick list" for advertising offers so that be will automatically be informed of products that interest him. (Jurgen 1996: 52)

The sketch succinctly captures both the luxury consumption habits and refined tastes of the intended listener as well as the promise of a technology that delivers, not just hi-fi sound, but information and entertainment that was ubiquitous, customisable and accessible. In market research for consumer digital radio products, over a third declared that they would be prepared to pay a premium for high quality tuners designed for hi-fi systems (Tuttlebee and Hawkins 1998: 265). Arcam, the renowned British manufacturer of hi-fi equiment, became one of the first companies to develop audiophile quality DAB receivers and extolled its potential to extend audiophile listening experience:

> Digital radio is one of the most exciting developments ever in radio. The crystal clear sound, utterly silent background and interference free reception delivers a new level of perfomance from broadcast sources.
>
> The experience of a digital broadcast of a live symphony orchestra concert is astonishing. The sound quality can transport you to the event without having to leave the comfort of your favourite armchair – truly the best seat in the house ... We believe that this technology will transform the way we listen to radio ... With Digital Radio, the listener can concentrate on the performance, knowing the sound quality will never vary. (Arcam 1998)

The New York Times welcomed the new digital revolution as the 'biggest technological leap since FM technology was developed in the 1940s and 50s', offering the same 'high quality of sound – free of static and hiss – as the digital compact disks now replacing analog phonograph records' (Pollack 1990). Similar positive reviews in the press echoed this claim that DAB had made a significant breakthrough in the quality of radio receiver technology, comparing it favourably to the benefits of CD audio listening:

> DAB – or Digital Radio as it is now to be known – is a joy to listen to. Certain sections of the HiFi press have started to 'knock' the system but, in my experience, it is better than anything that has gone before. It's quiet, has good imaging, good transients and lays bare all the faults in the source material! Listen to a well balanced live concert, however, and all the best qualities will be apparent. There's an added bonus – the Radio 3 feed is free of the dynamics processing which is applied to the analogue services. The aforementioned well balanced live concert, and the records of course, have a much more natural dynamic. A very enjoyable experience. (Stokes 1998)

The Digital Dream

DAB, the technical literature tells us, when used to its full potential and presented under optimal conditions, does offer 'near CD-quality' and a detailed, high fidelity listening experience with little need of the dynamic compression that is frequently used for FM and AM transmission (Spikofski and Klar 2003). CD-quality has indeed become the popular standard by which to measure listening quality, even if its own claims to audio fidelity have been questioned (Rothenbuhler and Peters 1997). The adoption of the Compact Disc format in the period from the 1980s on as the audio benchmark (Josse 2002) represents the culmination of an extensive, complex and sometimes controversial history of audio technology development of which DAB is a part. Nothing short of the full digitalization of the audio chain was the over-riding goal of audio technology development in this period, resulting in inter-related professional and consumer audio innovations, including new digital compression standards, new transmission systems and different platforms for audio distribution and delivery.

Released in 1982, the CD audio standard was the result of a long term collaboration between Sony and Phillips to provide a replacement for the LP, or long playing vinyl format, that had dominated the music industry and recorded music consumption since the 1950s (Immink 1998; 2007). The compact disc introduced a number of radically different elements into audio technology: firstly, the use of digital data to store and process the audio signal; secondly, error correction to make the signal robust; and thirdly, the use of optical, non-contact pick-up to read the signal on the disc (Pohlman 1992: 8) all of which had the effect of overturning nearly a century of analogue audio evolution and setting an entirely different trajectory for its future development. In parallel with the development of analogue audio recording techniques, a series of milestones in digital audio includes the development of digital audio sampling in 1928, the development of pulse code modulation in 1937, and further extensive related developments in computing and digital signal processing in the post second world war period. Working digital audio recording systems were demonstrated by Sony and Japanese broadcaster NHK in 1969 and by the early 1970s the BBC was using digital recorders for master recording (Pohlman 1992: 10). Drawing on their respective experiments with different forms of optical storage for audio and video content, including the Laservision system, Sony and Philips agreed in 1979 to collaborate on the design of the compact disc system, and formally introduced it to Europe and Japan in 1982.

The perfectability of audio recording and reproduction through digital means may be seen as a further dimension of what Vincent Mosco (2005) has referred to as the 'digital sublime' – a mythic belief in the power of new digital technology to open up a new world of possibilities. This unswerving belief in the power of engineering and scientific progress, the 'technological sublime' reviewed by David Nye (1994) and earlier by Leo Marx (1964), was captured in CD's marketing slogan, 'Perfect Sound Forever', conveying all the hype and the exaggeration attached to what was at best a compromised technical solution. The compact disc format, now over twenty-five years old, has never been wholly accepted among some audiophiles or audio purists, and was a source of major frustration for those who found its sound to be 'clinical' or 'harsh' compared to the analogue 'warmth' and musicality of older analogue technology (Harley 1998: 255). In the audio world, a schism was effectively created between diehards who believed only in analogue methods and proponents of digital audio fidelity (see Rothenbuhler and Peters 1997; Kessler and Harris 2005: 207). Such disputes aside, the success of the CD in entirely displacing, from the late 1980s on, the distribution of pre-recorded music on LP vinyl and audio cassette is indisputable and remarkable (Hansman et al. 1999; Goode 2002). Later enhancements to high end CD audio technology, as well as development of next generation high definition digital audio formats such as Super Audio CD (Aarts et al. 2004), have retained the utopian goal of perfect digital audio reproduction for the consumer market.

Against this background, the audio technology developed as part of the DAB standard can be seen as part of a general movement towards making the benefits of digital audio available across the full entertainment spectrum, in this case within the broadcasting transmission chain, and to complete a process of digitalization that had been well established in all other aspects of audio recording, storage and reproduction. Given the large quantities of data involved in CD audio sampling (16 bit audio sampled at a frequency of 44.1 KHz), methods of compression and reduction of the amount of data to be transmitted were essential to making digital audio transmission systems possible. Some of the most important work on audio compression standards and on encoding and decoding audio signals was undertaken within the Eureka 147 consortium, in particular by the Institut für Rundfunktechnik (IRT) and by the Fraunhofer Institute, in developments that led ultimately to the development of the MP-3 algorithm (Musmann 2006; Sterne 2006; Fraunhofer IIS 2008). Pursuing a goal of being able to transmit high quality digital audio over ISDN lines, researchers at the Fraunhofer Institute developed a number of encoding schemes to make digital audio transmission more manageable. The Moving Picture Experts Group (MPEG), established in 1988 as a working group of the International Standardisation Organisation (ISO) to assess compression standards for digital audio and video, approved in 1992 the MPEG-1 compression standard, comprising three distinct components (Layers 1, 2 and 3). The more complex Layer 3, to be later known as MP-3, was first used for professional applications in radio stations and studios for ISDN transmission, but laid the foundations for a global revolution of music storage on PC and portable media players, transmission over the Internet and in peer-to-peer file sharing. The somewhat less complex Layer 2, or MP-2, was selected by DAB as the audio format for digital audio broadcasting services. MP-2 was developed as a psycho-acoustic compression algorithm, or in other words, used analysis of human auditory perception capabilities and limitations to produce an efficient means of removing the unnecessary parts of the audio signal. The claim for both MP-2 and MP-3 was to produce near CD-quality at greatly reduced data levels.

DAB was not the only service heralding the sound of the future using the newly developed digital technologies of compression and transmission in the early 1990s. The concept of digital music transmission over satellite and terrestrial systems developed strong industry currency, and a number of major initiatives in addition to DAB were proposed. Firstly, as a precursor to later satellite radio services, companies such as Digital Cable Radio and the Digital Music Company in the United States and CBC's Galaxy service in Canada began to offer subscription-based digital music services via cable television lines, with multiple channels offering different genres of music services and simulcasts of major pay-per-listen events, such as headline concerts and sporting events (Pollack 1990; Walker 1991). At the same time, proposals for satellite transmission to both home and in-car receivers were also being actively developed. Sirius Satellite Radio, now merged with its erstwhile rival XM Radio, began life in 1990 as Satellite CD Radio and proposed a digital radio service that would be broadcast by satellites to listeners with special receivers, earning its revenues from charging subscription fees. Based on the idea that the future of music and audio entertainment would be driven by the near unlimited choice through multiple channels, CD-quality audio, and exclusive contracts with stars from the entertainment world and for certain sports events, Sirius and XM Satellite Radio subsequently invested billions of dollars in satellite technologies and programming available across North America and on the Internet. Such developments in the context of rapid digitalization and reconceptualisation of how audio services might adapt to new technological possibilities were likened at the time of their development to HDTV issues for radio: a transitional moment in which whole aspects of the service would be reconsidered, involving enormous potential disruption for the industry, but which were an inevitable and necessary stage in order to prosper in the digital future.

'Worse than FM'

Of the many claims for digital radio's supposed enhancements, none has been quite as controversial as that of its supposed CD-quality sound, and no other feature has attracted the same degree of ire and listener frustration as the audio quality of DAB. From a technical point of view, DAB digital radio sounds excellent when transmitted at the originally-envisaged bitrate of 320kbps (Spikofski and Klar 2003). More often than not, however, bit-rates are determined the minimum necessary for acceptable listening, not the maximum or even the recommended levels for effective audio performance. The fact that more content can be offered by reducing the bitrate per station on a digital mulitplex has meant that broadcasters (and consumers) have tended to prefer quantity over quality. Multiple bit streams and compression, therefore, are extended to the highest acceptable limit resulting in a quality of transmission that is frequently perceived as 'worse than FM'.

Critics of digital audio and digital radio have not been slow to air their views on the failings of DAB in this regard. Decrying the rising popularity of reduced bit-rate systems such as MP-2 or MP-3 which use compression as the basis for storage or transmission, *Stereophile*, the leading US audiophile journal, argued in 1992 that we might one day lament the passing of the 'golden age' of digital audio 'when consumer formats (CD and DAT) contained a bitstream that was an exact bit-for-bit duplicate of the original studio master recording – not a digitally compressed, filtered, copy-resistant version whose sound is "close enough" to the original' (Mitchell 1992). Compression, the magazine's editorial continued, is antithetical to the spirit of high end audio and has more to do with practical and economic goals – making recorded signals available to consumers in formats that are more compact, portable or affordable. The difficulty with glossy compression algorithms based on psycho-acoustic perceptual coding, it was argued, is that while they economize on data streams by eliminating that part of the audio which most listeners won't actually hear, this is an approximation for 'some listeners in most situations'. The compact disc in this sense represents the limits of compromise between a mass consumer format whose signal delivery can also satisfy critical listeners. Reduced bit rate schemes such as MP-3 for audio purists are simply unacceptable and in all cases exhibit a noticeable deterioration in audio quality. Comparing a satellite radio feed to the original CD source, one *Stereophile* reviewer commented:

> The MP3 sounded flat, dull, and lifeless in comparison … The sense of air and space was lost, and, consequently, the emotional impact was drowned … Even the performance suffered – the singers sounded as if they had become tired and sloppy. (Atkinson 2008)

A vocal albeit minority audience response to issues of the audio quality of DAB transmissions prompted regulators and other agencies to investigate complaints about its performance, as well as to assess how important quality of audio was to listeners. Technical assessments of the audio quality are difficult to achieve, and are normally based on a combination of objective and subjective tests of programme loudness levels and perceived audio quality. A study of German DAB transmission quality found that, despite high bitrates, broadcasters frequently employed audio processing in inappropriate ways – usually in attempt to boost the signal's apparent loudness over its competitors – that greatly degraded the audio signal with little concern for music fidelity (Spikofski and Klar 2003).

In the United Kingdom, a lack of available frequencies, combined with a general explosion of media choice, also resulted in a reduction of bitrate per station to add more choice, with a consequent reduction in audio quality. In some instances this has led to an outcry about the poor quality of DAB. An article by Jack Schofield, for instance, published in *The Guardian*'s Technology section (23 November 2006 "The future of UK radio is now in your hands") was severely critical of DAB's audio quality and of the future for the standard. In it, he criticised the outdated and inefficient MP-2, arguing that audio coding had progressed through MP-3, AAC and now AAC+. Accordingly, particularly given that other countries had rejected DAB in favour of new better standards such as DAB+, the United Kingdom should now follow likewise. Extensive discussion of DAB quality issues became a recurring theme of BBC discussion programmes, the quality press and the blogosphere, in which a variety of known performance issues have been highlighted and criticised. The UK blog *Digital Radio Tech*, for instance, continued the audiophile theme of unacceptable DAB audio, listing its shortcomings as:

- Very poor top-end (high-frequency) response (because using 128kbps instead of the preferred 192kbps cuts off the higher frequencies)
- Dull sound due to the poor top-end response
- Muffled sound due to the lack of accuracy at which the audio samples are encoded at due to the low bit rates used
- Lack of stereo image and instruments all meld together to form a messy, muddled "wall of sound"
- Swishy vocals
- Sibilant speech (when people pronounce 'ss' or 'sh' sounds they come out sounding 'lispy').
(Digital Radio Tech 2006)

Respondents to the blog echoed all of the above and more, one correspondent commenting that:

To my ears the DAB sound is plain awful. Harsh and tinged with low level hash that makes listening fatiguing. I have sampled a few DAB radios and they all have this nasty quality ... So if Apple can achieve very aceptable results with 128kb audio why not DAB? (Digital Radio Tech 2006)

Such vocal public criticism prompted Ofcom, the UK regulator, to independently assess the extent to which members of the public were concerned about the apparent poor audio performance of DAB broadcasts. In its consultation for *The Future of Radio,* over 70% of responses had in some way questioned the audio quality of DAB broadcasting, raising, among other issues, that as long as DAB appeared to be inferior to existing FM services, there could be no question of a switch-off of analogue radio transmission (Ofcom 2007). Respondents also queried what appeared to be the retrograde step of broadcasting some DAB services in mono (when stereo FM was the norm), as well as arguing for the adoption of better codecs such as AAC and DAB+ (Ofcom 2007: 113). Ofcom consequently undertook its own independent research into consumer perceptions to ascertain whether there was widespread dissatisfaction with the quality of DAB transmission, or whether it was confined to a small number of audiophiles. The research found that for the vast majority (over 80%), the sound quality of DAB was excellent or very good, a response that was the same for both regular DAB listeners as well as those who were also Hi-Fi owners. Respondents were also asked to rate DAB sound quality compared with FM, and similarly high numbers supported the view that DAB sounded at least as good or better. 94% of all DAB listeners said it was at least as good as FM, with 77% saying it was better than FM. Only 3% thought it was worse than FM.

The Ofcom research concluded that there was little evidence that the majority of the public supported the view that DAB audio quality standards were inferior to FM or had deteriorated to the extent claimed by audiophiles. This, the regulator added, was not to say that such criticisms are wrong or misplaced; it was simply that their expectations of audio standards were not shared by the vast majority of listeners (Ofcom 2007: 115). On the face of it, the vocal minority of audiophile critics was just that – a minority – and in the trade off between capacity and audio quality, the provision of new services at lower but acceptable bit-rates had been deemed to be the better strategy. With respect to the issue of the adoption of newer compression technologies in DAB+, the Ofcom strategy document observed that the implementation of AAC coding did not on its own imply improved sound quality. Given that sound quality is a function of the bit-rate used by the broadcaster to transmit the signal, DAB+ would add extra capacity but ultimately the broadcaster (or multiplex operator) would have to 'make a trade-off between the number of services (audio or data) fitted in to the multiplex and the sound quality of those services' (Ofcom 2007: 115).

Conclusion

DAB's 'sound of the future' promise was a bold and ambitious declaration produced by the first promoters of Eureka 147 keen to establish a new vision for radio in the latter part of the twentieth century. The utopian promise of a 'perfect sound machine' to match the marketing slogan for CD, 'Perfect Sound Forever', was formulated in an historical context when digitalization appeared both to resolve technical issues or constraints found in older analogue processes, and to herald new, more creative and interactive possibilities of making content available. The experience of DAB, as discussed in this chapter, suggests that such claims were not necessarily wrong or invalid within the context within which DAB evolved and developed; rather that its implementation alongside related developments in digital audio technology created a range of possibilities, including new and previously unimagined ones, that conflicted with the original founding vision of what DAB Eureka 147 was trying to achieve.

Among the many unintended consequences of the development of efficient compression technologies and means of distributing digital audio, was the ease with which digital audio files of acceptable and 'good enough' quality could be stored, streamed and shared on the Internet, with momentous consequences for the music industry and for the social production and consumption of recorded music. The fact that DAB digital radio was effectively caught in the cross-fire of a massive reorganisation of the music industry distribution model, with an extended preoccupation with management of digital rights, did not assist the case of DAB's inherent potential for higher quality audio transmission.

As industry sources quote, the experience of market implementation of DAB wherever it has been launched is to the effect that listeners wanted new additional services and a new value proposition to make an investment in digital radio. Audio quality on its own was not, as the original developers envisaged, sufficient to encourage interest of market take-up. In an era when new media services offered extensive choice and capacity, the primary focus for successful digital radio implementation, to the exclusion of maximising audio quality, was on providing additional content, new channels and services unavailable on traditional analogue radio. The primacy of the audio experience was clearly no longer the focal point of attention and had to take second place alongside rival competing services available online, and for personalised mobile media consumption.

Against this background, the resilience of FM radio as a robust and reliable medium can be better understood. As in the case of the replacement of the analogue LP vinyl format by CD, FM has proven to be much more resilient than originally imagined, and analogue switch off is not a realistic prospect in any existing market at this time. In response, the industry has sought to provide added value and a new dimension to digital radio offerings by suggesting a new era of high resolution, downloadable formats, and multichannel surround sound transmissions and downloads. Given that it is only a question of bandwidth and storage capacity that sets limits on today's compressed, low-resolution audio, companies such as DTS have launched experimental broadcasts and webacasts looking ahead to high quality audio multicasting over broadband, particularly once fast connections over 8Mbs become more commonplace (Iverson 2004; Barbour 2005). Whether this proves to be a qualitative leap forward and a transformational paradigm for digital radio in the next decade, or another false dawn, remains to be determined.

References

Aarts, R. M., D. Reefman and E. Janssen (2004), 'Super audio CD-an overview', Eindhoven: Philips Research Labs, www.extra.research.philips.com/hera/people/aarts/papers/aar04pu3.pdf. Accessed 26 May 2009.

Arcam (1998), 'Product Description: Alpha 10 Tuner', http://www.arcam.co.uk/alpha/tuners/alpha10. html. Accessed 14 November 2008.

Atkinson, J. (2008), '"CD-Quality": Where Did the Music Go?', Stereophile, February 2008.

Barbour, J. (2005), 'Multi-Channel Surround Sound on Digital Radio', Radio in the World: Papers from the 2005 Melbourne Radio Conference, Melbourne, Vic: RMIT.

Digital Radio Tech (2006), 'DAB sound quality article on You & Yours', http://www.digitalradiotech. co.uk/articles/DAB-sound-quality-article-on-You-amp-Yours.php. Accessed 21 November 2008.

Fox, B. (1991), 'Radio sans frontieres: By the mid-1990s, people driving across Europe should be able to tune into their favourite radio programmes in hi-fi wherever they are', *New Scientist*, 20 July 1991, p. 29.

Fox, B. (1994), 'The perfect sound machine', *The Times*, London: Times Newspapers Limited, 14 October 1994.

Fraunhofer IIS (2008), 'The mp3 history', http://www.iis.fraunhofer.de:80/EN/bf/amm/mp3history/ mp3history01.jsp. Accessed 26 May 2009.

Goode, M. M. H. (2002), 'Predicting consumer satisfaction from CD players', *Journal Of Consumer Behaviour*, 1: 4, pp. 323–335.

Hansman, H., C. H. Mulder and R. Verhoeff (1999), 'The Adoption of the Compact Disk Player: An Event History Analysis for the Netherlands', *Journal of Cultural Economics*, 23: 3, pp. 221–232.

Harley, R. (1998), *The Complete Guide to High End Audio*, Albuquerque, NM: Acapella Publishing.

Hoeg, W. and T. Lauterbach (2001), *Digital Audio Broadcasting Principles and Applications of Digital Radio*, Chichester: John Wiley & Sons.

Immink, K. A. S. (1998), 'The Compact Disc Story', *Journal of the AES*, 46, pp. 458–465.

Immink, K. A. S. (2007), 'Shannon, Beethoven, and the Compact Disc', *IEEE Information Theory Newsletter*, December, pp. 42–46.

Iverson, J. (2004), '5.1 96/24 Audio Downloads?', *Stereophile*, 26 July.

Josse, D. (2002), 'DAB – now hitting the market on an industrial scale', *EBU Technical Review*, 292.

Jurgen, R. K. (1996), 'Broadcasting with digital audio', *IEEE Spectrum*, 33: 3, pp. 52–59.

Kessler, K. and S. Harris (2005), *Sound Bites: 50 Years of Hi-Fi News*, London: IPC Media.

Lau, A. and W. F. Williams (1992), 'Service planning for terrestrial Digital Audio Broadcasting', *EBU Technical Review*, 252.

Marx, L. (1964), *The machine in the garden: technology and the pastoral ideal in America*, New York: Oxford University Press.

Mitchell, P. W. (1992), 'A Question of Bits', *Stereophile*, September.

Mosco, V. (2005), *The Digital Sublime: Myth, Power, and Cyberspace*, Boston, MA: The MIT Press.

Musmann, H. G. (2006), 'Genesis of the MP3 audio coding standard', *IEEE Transactions on Consumer Electronics*, 52: 3, pp. 1043–1049.

Nye, D. E. (1994), *American technological sublime*, Cambridge, MA. and London: MIT Press.

Ofcom (2007), *The Future of Radio – The next phase*, London: Office of Communications.

Ofcom (2007), *The Future of Radio: The future of FM and AM services and the alignment of analogue and digital regulation*, London: Office of Communications.

Pohlman, K. C. (1992), *The compact disc handbook*. Madison, WI: AR Editions Inc.

Pollack, A. (1990), 'Next, Digital Radio For a Superior Sound', *New York Times*, 11 July 1990.

Rothenbuhler, E. W. and J. D. Peters (1997), 'Defining Phonography: An Experiment in Theory', *The Musical Quarterly*, 81: 2, pp. 242–264.

Shelswell, P., C. Gandy, J.L. Riley and M. Maddocks (1991), 'Digital Audio Broadcasting', *IEE Colloquium on Vehicle Audio Systems*, London, UK.

Spikofski, G. and S. Klar (2003), 'DAB and CD quality – reality or illusion', *EBU Technical Review*, 296.

Sterne, J. (2006), 'The mp3 as cultural artifact', *New Media Society*, 8: 5, pp. 825–842.

Stokes, A. (1998), 'The Arcam Alpha Ten DAB Tuner', http://www.mb21.co.uk/ether.net/radio/ alpha10.shtml. Accessed 14 November 2008.

Task Force on Digital Radio (1995), *The sound of the future: The Canadian vision*, Ottawa, ON: Minister of Supply & Service.

Tuttlebee, W. H. W. and D. A. Hawkins (1998), 'Consumer digital radio: from concept to reality', *Electronics & Communication Engineering Journal*, 10: 6, pp. 263–276.

Walker, K. (1991), 'Digital Audio Broadcasting: Radio Wave of the Future', http://www.exhibitresearch.com/kevin/media/dab.html. Accessed 20 December 2007.

Chapter 5

'DAB: The Future of Radio?' The Development of Digital Radio in Europe[1]

Per Jauert, Stephen Lax, Helen Shaw & Marko Ala-Fossi

D igital radio reached its fifteenth birthday in 2010. However, anniversary celebrations have been rather muted. In contrast with digital television, where audience awareness levels are high and adoption rates similar, digital radio's penetration into the marketplace has been minimal. Even when people have heard of digital radio, survey data shows that very few feel knowledgeable about it.

Nevertheless, most broadcasters and most governments believe that just as most other consumer electronics are migrating from analogue to digital technology, radio too will eventually become all-digital. But development of digital radio is very variable, with some countries having a large number of digital radio services available across the whole nation, while others have very few services.

The picture is confused by there being a number of interpretations of the term 'digital radio'. Almost all households that receive digital TV, whether by satellite, cable or over-the-air terrestrial transmissions, will also receive a number of radio stations with that service. So listening to a radio station through the TV is one form of digital radio service. A second means is to listen to streamed radio stations on the Internet – this too can be regarded as digital radio. However, one of the key attributes of 'traditional' analogue radio is the ability to receive it in a number of locations, perhaps on a number of different portable radios located around one's house, or on a personal stereo (such as the 'Walkman'), or in a car. None of the existing digital television systems allow for this portability or mobile reception; it is also currently not practicable to listen to web-streamed radio on portable devices. Instead dedicated digital radio systems have been developed which replicate all the attributes of analogue radio in digital form, and the most advanced of these is the terrestrial DAB system (for Digital Audio Broadcasting), sometimes known as Eureka after the EU research programme under which it was developed. The development of DAB began in 1988 with the involvement of broadcasters and equipment manufacturers from four countries, and domestic transmissions began in September 1995. Ten years later, 28 countries were operating DAB services (World DAB 2005), mostly in Europe but also including some advanced services in Canada and trial services in Australia, China and South Africa. Notable absentees from the list are the United States and Japan which have opted for different systems.

DAB was described from the outset as a potential replacement for analogue FM radio. Like FM, it would offer high quality sound (in comparison with AM) with a range of both national and local or regional stations. The advantages digitization would offer over and above FM included a greater number of stations, easier tuning of radio sets and the display of text services on the receiver display. The expectation was that while DAB would remain essentially a technology for the delivery of radio services, the data carried on the DAB transmissions could also include multimedia information, and that the radio would become a more sophisticated device capable of receiving graphical information and with the ability to store and replay broadcast audio. The digitization of radio then presented a number of new possibilities for augmenting the service and even challenging the meaning of the term 'radio'.

However, these advantages have not proved compelling to listeners. Sales of DAB receivers have been slow in most countries in Europe with the exception of Denmark, the UK and Norway. Although they are now cheaper than a few years ago, they continue to carry a premium in price over comparable analogue receivers, and thus listeners need a clear incentive in order to make the decision to purchase. This slow growth has also made some broadcasters and governments wary of investing in the DAB system. DAB requires the allocation of new frequency space and a reorganisation of the broadcasting system, from one based on allocating particular frequency channels to individual stations, to one in which a wide frequency channel is allocated to a 'multiplex operator', which then carries a number of radio stations simultaneously on that channel. Thus, there are legislative and regulatory processes involved. For the broadcaster, the introduction of DAB requires the conversion of its transmitter network to simulcast existing services, or the negotiation for the carriage of such services with the new multiplex operators. At the same time, the increased capacity of DAB implies that those same services will be competing with a host of new stations carried on the various multiplexes. Hence, depending on the particularities of the existing radio broadcasting landscape, from country to country there may be varying degrees of incentive for broadcasters and their governments to embrace DAB digital radio.

Thus, while some of the early visions for digital radio may have been cautious, even they now appear to have been over optimistic. Quentin Howard, head of the UK digital commercial operator Digital One, argued in 1999: 'All media is going digital... To think that analogue radio will still be able to hold its own in ten years' time is unrealistic: it won't be able to compete' (quoted in Carter 1999: 46). Now that we have reached that horizon, it seems clear that analogue radio is indeed holding its own and will continue to do so for many years to come. Few countries have discussed switching off analogue radio and fewer still have actually specified a date for doing so. Once again, the contrast with television is striking with several countries expecting to switch off analogue TV within the next decade. In attempting to explain the reasons for the relatively slow development of digital radio, the large degree of variation between different countries may well offer some clues.

The level of development of DAB services varies greatly: from none at all to relatively advanced. While, as we have noted, there are some developments outside Europe, in some cases based on quite different technical standards, it is in Europe where the most developed services are to be found. The DAB project began in Europe, and it is most advanced in northern Europe where the project's member countries are principally located. The level of DAB service can be judged on three criteria: the geographical coverage of DAB signals; the number of stations available on DAB; and the number of receivers in use. On this basis, some countries can be considered to have well developed services: for example, with wide geographic coverage, public (and some commercial) stations on digital, and a significant adoption of receivers by the public. Others might have a medium level of development, with some services operating but far from national geographical coverage and a consequent low take up of receivers. Low levels of development exist in other countries with often only trial DAB services and almost no public adoption of receivers. Some examples are listed in Table 1.

Necessarily, the picture illustrated in Table 1 represents a snapshot. Particularly in the early days of its introduction, the level of service in any particular country can change rapidly, and so recording the state of development at a particular time gives only a partial

Table 1: DAB after fifteen years: levels of development of selected European countries in 2008–2009. Data extracted from Public Radio in Europe 2007, World DMB 2009 and national statistics.

High levels of DAB services

Denmark
Launch: 2002
Coverage: 90%
Channels: 15 and 3 private radio channels on DAB
Receivers: 1.3 m
Penetration by household: 30 %

Germany
Launch: 1998
Coverage: 85% coverage.
Channels: 45 public and 62 private channels on DAB
Receivers: 546.000
Penetration by household: na

United Kingdom
Launch: 1995
Coverage: 85%
Channels: 45 public channels and 172 private channels
Receivers: 8.5 m
Penetration by household: 20 %

High to medium levels of DAB services

Belgium
Launch: 1997
Coverage: 100%
Channels: 9 public channels on DAB
Receivers: na
Penetration by household: na

Norway   Launch: 1995
                            Coverage: 80%
                            Channels: 11 public and 2 private channels on DAB
                            Receivers: 290.000
                            Penetration by households: 15%

The Netherlands   Launch: 2004
                            Coverage: 70%
                            Channels: 9 public channels on DAB, 3 on DRM
                            Receivers: na
                            Penetration by household: na

Medium to low levels of DAB services

Ireland   Launch: 2006
                            Coverage: 44%
                            Channels: 10 public, 7 commercial (on trial in 2007)
                            Receivers: na
                            Penetration by household: na

Sweden   Launch: 1995
                            Coverage: 35% (Reduced from 85% in 2001)
                            Channels: 7 public channels on DAB
                            Receivers: na
                            Penetration by household: na

Low levels of DAB services

Finland   Launch: 1998 – ended 2005
                            Coverage: 40%
                            Channels: 8 public
                            Receivers: na
                            Penetration by household: na

France   Launch: 2007–2008 (trial only)
                            Coverage: 50%
                            Channels: 8 public channels on DAB
                            Receivers: na
                            Penetration by household: na

Spain   Launch: 1998
                            Coverage: 52%
                            Channels: 6 public channels, 6 private channels on DAB
                            Receivers: na
                            Penetration by household: na

understanding of DAB's status. Instead, a fuller understanding is obtained by looking at the development over time of DAB in different countries, and identifying the different factors which have influenced that development.

We primarily examine here the long-term development of DAB in mainly four countries from Table 1: the United Kingdom, Denmark, Finland and Ireland, but will also include some central elements from other European countries, i.e. France, Germany, Norway, Spain and Sweden. These countries straddle the full range of levels of DAB's development across Europe and allow us to examine the role of governments, broadcasters and the audience in shaping the development of digital radio. The United Kingdom is recognised as the leading country in Europe (and indeed in the world) in terms of the number of stations broadcast on DAB, both public and commercial, and in terms of the number of receivers in households. Denmark is also relatively advanced in having all public stations on DAB, including some new digital-only stations, although commercial radio has begun DAB transmissions two or three years later than DR. Measured by penetration by households, Denmark overtakes the UK. Finland has operated a DAB service for some years, like Denmark carrying only the public broadcaster's stations. However, take-up of receivers by the public has remained low, and the public broadcaster switched off its transmissions in August 2005, with fewer than 1000 receivers sold by 2004 (YLE 2004). Ireland has had a slow start to digital radio implementation. Following initial domestic trials, there was a hiatus in which no service was offered. A further trial in 2006 was followed by limited digital radio services offered by the public broadcaster in the three main urban centres of population in the country. No commercial services are available on the DAB network. The different approaches in these four countries demonstrate the complexity of the processes involved in introducing a new digital technological system into broadcasting, and the limitations of an approach based largely on technical assumptions adopted in the initial stages of DAB's development.

DAB – four brief histories

The Eureka 147 project under which DAB developed reached fruition in the mid-1990s when a number of European countries began trial broadcasts, with full domestic services beginning soon after. Public service broadcasters were the first to offer digital services: in the UK, the BBC transmitted its five existing stations in digital format from 1995, while YLE in Finland initially began digital-only services from 1998. In Denmark, DR began full domestic DAB services in 2002. In those early days, coverage levels were low, typically reaching around 40% of each country's population, but in the UK and Denmark coverage grew steadily as more transmitters were built (or existing ones converted), and in both countries 85% or more of the population can receive DAB signals. As we have noted, in Ireland, no domestic DAB service emerged after the initial trials, which ceased in 1999, although RTÉ recommended further, Dublin-based trials in 2006 and launched a limited service in 2008.

Frequencies for DAB were allocated to each country at the Wiesbaden World Radio Conference in 1995. Each frequency block carries a single DAB multiplex, and each multiplex can be coded to transmit typically between five and ten radio stations. A multiplex can be relayed on a 'single frequency network' to provide national services to a whole country, or used on a regional and local basis, and most countries operate both national and local multiplexes in a manner similar to the allocation of analogue FM frequencies. Digital transmission differs from analogue though in that allocating a DAB multiplex to a particular area in effect creates capacity for around ten radio services in that locality. For national networks, this is not a particular problem. Public broadcasters in most cases already operated a suite of national stations: the BBC, YLE and DR were able to simply simulcast existing stations on a national DAB multiplex, with the option of adding new, digital-only services later. For local services, however, the capacity of the new DAB multiplex would often exceed the number of existing local radio stations operating in that area. Thus, the introduction of a DAB multiplex structure created a potential dramatic increase in capacity for radio stations, both local and national. At least one commentator suggested, perhaps over-flamboyantly, that the introduction of digital transmission would 'probably mark the first time in broadcasting that there will be more channels available than content to fill them' (Crisell 2002: 279).

In all four countries considered here, radio is strongly regulated with established public service policies, and it is to be expected that policy for digital radio would see a leading role for each country's respective public broadcaster. The expectation was that the public broadcasters would build the first networks, simulcasting their existing suites of stations, and that in time commercial radio companies would follow this lead. The steps taken to encourage commercial radio to invest in digital radio reveal a marked difference in policy decisions taken by the different countries. In the UK, while one national multiplex was operated by the BBC, the additional national multiplex and all the local or regional multiplexes were to be operated by commercial companies, which would in turn contract with commercial radio companies in order to carry their stations, whether national or local. In most cases the fees charged to those stations for digital transmission were higher than the costs of analogue transmission, and so represented a significant cost. However, under the provisions of the 1996 Broadcasting Act any existing analogue commercial radio station which began digital transmissions would be granted automatic renewal of its analogue broadcasting licence, a valuable commercial asset and thus a significant incentive to invest in DAB.

Elsewhere, in both Finland and Denmark, no equivalent incentive was offered to commercial radio, but instead a condition of the national stations' analogue licences was a requirement to begin digital transmissions at some later date (local commercial stations were not expected to join DAB at this stage). Here the plan was to use the public service broadcasters as the DAB locomotive. DR and YLE were charged with developing and operating the DAB networks in order to develop the market. Once established, or at least demonstrated with sufficient numbers of receivers sold, the expectation was that commercial radio would willingly join in DAB's further development. In Ireland, where there remains unused capacity within the analogue FM spectrum, commercial radio companies showed little interest in digital radio and thus the public broadcaster RTÉ, who was responsible for initial trial transmissions, also saw little reason to continue after that.

The expectation that, other than in Ireland, commercial radio would enthusiastically embrace DAB has not proved to be the case. In Finland, on renewing its analogue licence in 2001, the national commercial station Radio Nova successfully negotiated away its obligation to transmit on digital. In Denmark, the two national commercial stations, Sky Radio and Radio 100FM, only began simulcasting in August 2005, though rather reluctantly as both companies expected to lose money on their digital services. In fact, three months later one of the stations, Sky Radio, ended all broadcasts – analogue and digital – for financial reasons, the costs of DAB transmissions being only one, relatively minor factor. In the UK, while commercial radio does have a significant presence on DAB, nevertheless more than half had not begun digital transmission by Autumn 2005, principally on economic grounds: the cost was simply too high. In many cases these were the smaller stations which were not part of the large radio groups, and which were therefore less able to afford the investment.

The presence or absence of commercial radio on the DAB multiplexes, on the basis of these four cases, would appear to be an indicator of the general health of DAB in these countries. In the UK, in particular, and more recently Denmark, each have a level of involvement from commercial radio, backing up the initial efforts of their public service broadcasters. In comparison with other countries, comparatively high numbers of receivers have been sold (see Table 1), although these sales figures still represent only a minority of households. In Finland however, commercial stations have not begun DAB transmissions and, correspondingly, public broadcaster YLE clearly saw little future in the technology having already switched off its DAB transmissions in August 2005. Instead, YLE has offered digital broadcast radio services in the nationwide DVB-T network where two private radio companies are also engaged (Ala-Fossi and Jauert 2006: 71). In addition, the two largest commercial radio companies in Finland have started digital broadcasting in the new DVB-H-network for mobile television, but the popularity and use of these services has remained marginal (Lassila 2008). In Ireland, while some commercial stations participated in the most recent 2006–2008 trial, only the public broadcaster RTÉ is available on the DAB multiplex.

Hence, we see in these four countries considerable variation in the implementation of a digital successor to analogue FM radio. With broad similarities in their broadcasting histories this might have suggested that digital radio would also develop in similar ways, but this has clearly not been the case. We focus next on a number of other European countries where digital radio has had a presence, before proceeding to analyse the differences between policies adopted by the government in each country, and seek to explain these variations by examining the existing radio 'landscapes' which these policy decisions reflected.

Perspectives from other European countries

Sweden and Norway were among the first European countries to launch DAB but the outcome for the two countries has been quite different. The Swedish government and the public service broadcaster Swedish Radio (SR), as well as the delivery company Teracom, were all probably rather convinced about the success of digital radio because already by 1999 the DAB transmitter network was expanded to cover 85% of the Swedish population. In spite of the early introduction of seven DAB-only channels, DAB audiences remained almost non-existent. At the end of 2001, SR and Teracom finally decided to call time and shut down most of the DAB network. Since then DAB services have only been available in four main metropolitan regions and the coverage was reduced from 85% to 35% (Ala-Fossi and Jauert 2006: 78; SOU 2002: 38). On the edge of a final shut down of DAB like in Finland, the Swedish government in 2006 commissioned the Swedish Radio and TV Authority to monitor the development of digital radio in Sweden on the basis of an ongoing assessment of different technologies, with the report from the Authority published in 2008 (Swedish Radio and TV Authority 2008). On the basis of audience surveys that showed a 45% population support for more channel diversity and new radio features, the report recommended a re-launch of nationwide digital radio coverage, based on the DAB+ standard. The report was supported both by the public service broadcaster SR and the main commercial players on the analogue local radio market, MTG and SBS.

Like Sweden, Norway launched DAB in 1995 but with a more modest level of ambition. NRK, the public service broadcaster, was the main player and like in many other European countries the commercial radio sector was quite hesitant to get involved in DAB. The DAB situation in Norway is in many ways comparable to the development in Denmark. The public broadcasters, supported by the government, share a common interest in stimulating the market, and have invested financial and human resources in the establishment of transmitters, in creating niche channels and in marketing efforts to stimulate the sale of DAB receivers. The official media policy and the legislation show differences between the two countries, but thus far the ministries in charge have not hesitated in keeping to the initial decision of introducing DAB as the main technology for digital radio broadcasting in contrast to the situation in Finland and Sweden (until 2008). In official documents, the shut down of the FM band in Norway has been set for 2015 but a final decision has not been taken yet, though the political majority claims that such a decision, comparable to digital television, will further stimulate the roll out of DAB.

In mid-Europe, Germany offers just as many public channels as the UK, and it reflects the large investments in programmes and technology by the 'Bundesländer' and the regional public broadcasters – in 2004 a total of €250 million (Kleinsteuber 2006: 138) – but it is also striking that the commercial radio sector, in contrast to the UK, has been more reluctant to engage in DAB. The structure of the German radio sector is quite different from other European countries. Firstly, the DAB regulation varies from region to region (Länder). Although the national regulator (RegTP) allocates the frequencies, content distribution for the commercial sector is licensed by the sixteen Länder media authorities. Secondly, Germany does not have a national public service broadcaster like the BBC in the UK, though ARD (Arbeitsgemeinschaft der öffentlich-rechtlichen Rundfunkanstalten der Bundesrepublik Deutschland) does broadcast in each of the sixteen Länder. Thirdly, there are eight DAB multiplex operators (one for each two Länder) and thus no single national multiplex. This might be one of the main reasons for the very low penetration. In spite of the high coverage and the high number of channels offered, only a very limited number of receivers (546,000) have been purchased (World DAB Forum 2009).

In March 2009, the sixteen German regional governments decided to start the roll out of a national multiplex using the DAB+ standard at the beginning of 2010, with capacity for ten to fifteen new channels. Deutschlandradio[2] will have access to five of these channels while the rest will made available to private operators. The national digital channels will be supplemented by two or three multiplexes in each of the sixteen regions. In total 30–40 digital radio stations (DAB and DAB+) will become available in every region of Germany, offering a mix of local and nationwide, private and public radio stations. In addition, Deutche Welle, Germany's international broadcaster, has engaged in DRM transmission on a trial basis since 2006 and plans an introduction of DRM+ in 2010 (EBU/UER 2007: 77–80).

In the Southern European region, Spain introduced DAB in 1998. In terms of territorial distribution, the autonomous regions of Spain each with its own regulatory authority have a certain resemblance to Germany's sixteen 'Länder', with 160 broadcasters and 225 channels (Bonet et al. 2009: 92). But compared to Germany, the Spanish radio landscape has a very different background and composition, and this may be one of the main reasons for the failure of DAB in Spain. In fact, DAB in Spain serves only as a secondary radio transmission service simulcasting FM channels. The only difference is that DAB has just a very few listeners (Bonet et al. 2008: 23).

France has tested a number of the digital radio standards over the years: DRM, DAB, DAB+, DVB-H, DVB-T, T-DMB and HD radio. The DAB tests took place in different regions of France, including some of the major cities, but in 2005 the development of DAB was brought to a halt, as the technology was not considered sufficiently good. The regulatory body for radio in France, Conseil Superieur Audiovisuel (CSA), decided in May 2009 to replace DAB with the T-DMB standard by 2008. The choice of standard was controversial and raised objections from the Digital Radio France organisation (DR France) because it 'not only reduces the number of broadcaster currently existing, but isolates France while closest neighbours have made different choices' (Hedges 2009).

Radio landscape and factors in the development of DAB

Radio in each of the four main countries studied here (Denmark, Finland, Ireland and the UK) follows what might be termed a classic Northern European model: a strong public service broadcaster maintained a monopoly of radio broadcasting long after public service television was joined by commercial competition. In radio, the first commercial stations were local, with national or regional coverage the domain of the public service stations (though public service, local radio stations already existed in the United Kingdom, provided by the BBC). In the United Kingdom, licensing of local commercial stations began in the 1970s, but accelerated after the 1990 Broadcasting Act, which allowed for the rapid expansion in the number of local commercial services, the launch of three national commercial services and a number of regional commercial stations. A similar pattern developed in Finland, where initial local commercial stations have been joined more recently by regional and national stations; in Denmark and Ireland, a third sector of much smaller, community stations developed in addition to local and regional commercial services, with one or two national commercial stations beginning in more recent years.

There have been significant differences in the *rates* of progress however, reflecting the sometimes precarious economic state of commercial radio. For example, Ireland's first national commercial station was launched in 1992 but failed early on, and a more careful management of licensed spectrum meant that it wasn't for another five years that a second attempt was made to provide national commercial radio, this time successfully, with Today FM remaining the only national commercial station in that country. In Denmark, it wasn't until 2003 that a national commercial station was licensed. The rapid growth of commercial radio in the United Kingdom during the 1990s almost exhausted the supply of FM frequencies and here, unlike other countries, the less valuable AM band (medium wave) remains extensively used for local and national commercial broadcasting. In Denmark too, expansion of commercial radio has left limited scope for expansion, and the commercial radio companies are arguing for spectrum reconfiguration to release further capacity inspired by the initial success of a similar initiative in the Netherlands. The regulators are hesitant due to the failure of the nationwide commercial radio sector, first introduced in 2003, and in part influenced by the rapid growth of the DAB roll out. After some years of hesitation, the shut down of the analogue FM band has again been an issue in the media policy debate as an alternative to spectrum reconfiguration since digital penetration is constantly rising, aiming for more than 30% in 2009. Finland also has only limited capacity remaining within the FM spectrum, but in Ireland the management of the spectrum has left a number of unused frequencies and the regulator, the Broadcast Commission of Ireland, announced a further programme of licences to be awarded from 2005.

Considering the 15 years of DAB activities in Europe, we can observe a number of similarities in the developments in the Northern European countries compared to Central and Southern Europe, although it seems obvious that the relative success of DAB is a Northern European phenomenon, while the inclination to maintain the digital modernisation of radio i.e., testing out formats other than DAB, seems to be most evident in Germany, and to a certain degree also in France. Only in Spain does it seem that a prospering of radio digitalization will have to wait for incentives for the commercial sector to be an active part of the process, perhaps on a DRM platform, which since 2007 has been in a test phase (EBU/UER 2007: 132).

So we can begin to explain the different approaches adopted by each of the four countries' governments: while there are certainly similarities between these countries' radio landscapes, there are also differences (see Table 2). These include: the availability of spare FM spectrum; the relationship between commercial and public service radio; and the balance between local and national services. These differences have all shaped the development of the DAB policy (and thus the current level of its development) in each country.

Table 2: The analogue radio landscape in Denmark, Finland, Ireland and the UK.

	Denmark	Finland	Ireland	UK
Geography:				
Population (million)	5.4	5.3	4.1	60.1
Area (km²)	43,094	338,145	70,273	244,101
Radio began:				
National	DR 1925	YLE 1926	RTÉ 1926	BBC 1922
Local/regional	1983	1976	1988	1967
Community	1983	1987	1988	2002
Commercial/private	1988	1985	1988	1973
National commercial	2003	1997	1992	1992
No. of stations:				
National public	4	4	4	5
Local/regional public	11	28	0	46
National commercial	2	1	1	3
Local/regional commercial	123	56	27	305
Local non-commercial/community	165	6	26	179
Advertising revenue:				
Radio's total ad revenue (€million)	33.5	50.5	97	623
Radio's share of all advertising (%)	2.0	3.4	9.0	5.9
Average daily listening (minutes)	131	195	244	191
Share of listening:				
All public stations	74	51	43	56
All commercial stations	26	49	57	42
Weekly reach:				
All radio stations	90	95	87	90
All public stations	87	61	54	66
All commercial stations	70	76	55	62

As the most advanced DAB country, the UK offers the clearest example of this shaping in its enticement of commercial radio. For, despite its rapid expansion resulting in a large number of stations, commercial radio's success here has been qualified. Taken as a whole, its share of the audience continues to struggle to match the BBC's, only once gaining more than 50%, and declining slowly to a low of 42% at the end of 2008 (RAJAR 2009). While its share of national advertising revenue grew during the 1990s when the number of stations was expanding, from 2004 its share has steadily declined (RAB 2009). With the limited availability of new FM spectrum, commercial radio could only anticipate incremental growth in analogue radio, but early decisions on digital radio policy enshrined in the 1996 Broadcasting Act made it clear that commercial radio was to have a leading role. With the awarding of the multiplex operating licences to the existing large commercial radio groups (when the first round of licence awards ended in 2003, the five major groups held all the licences between them, with additional smaller partners in some instances), this represented a significant shift in weight from the BBC towards commercial radio, particularly at the national level where there would now be as many commercial as public service stations. This emphasis was reinforced in 2007 with the allocation by the regulatory body, Ofcom, of frequencies for a second commercially-operated, national multiplex. While the additional capacity offered by DAB (approximately twice as many stations in any particular area than on analogue) would allow further expansion, the control of the multiplexes by the existing large radio groups coupled, as we shall see, with a relaxed regulatory regime, would help to give these groups a competitive advantage. The inclusion of a further particular incentive, the automatic renewal of existing analogue licences, finally persuaded commercial radio companies to risk the investment in DAB.

The contrast between the UK and Ireland is striking. In Ireland, the government avoided supporting any particular technical platform for digital radio and, following a report by consultants Deloitte and Touche in 2001, adopted a two-year 'wait and see' policy, awaiting evidence from other countries of the success (or otherwise) of DAB. This, again, can be explained by reference to the pre-existing state of analogue radio. Commercial radio here is mostly local (there are now just two national commercial stations), and in some cases *very* local. Meanwhile, the public broadcaster RTÉ has no local stations. Local radio has an unusually strong position in this country with some of the highest listening figures in Europe, and collectively has a powerful lobbying voice. There is also a healthy community radio sector. As noted earlier, the capacity exists within FM for further growth at both national and local level, though the experience of earlier failures in national commercial radio has resulted in caution on the part of the regulator in releasing spectrum, and meant only measured demand for expansion from the radio industry. Thus, the additional capacity offered by DAB was not a clear-cut benefit, and it was treated warily or even with hostility by the commercial radio stations. RTÉ itself was somewhat equivocal in its approach, and thus the government was under no particular pressure, economic or political, to accelerate the development of DAB. The local nature of many of Ireland's commercial stations, and the relative strength of the community radio sector, compounded the difficulties with the adoption of DAB – DAB technology is rather inflexible in comparison with analogue FM in its geographic coverage; while highly spectrum efficient for national coverage, DAB becomes far less efficient for smaller coverage areas. An indication of this indifference to DAB was given by the organisation representing Ireland's community stations, which argued that 'DAB is not a pressing issue for us at the moment,' suggesting that progress in the UK be monitored (CRAOL 2004). The government position has remained cautious: a second consultants' report in 2004 urged the establishment of a policy unit for digital platforms, but highlighted difficulties with DAB and suggested that 'the business case for DAB is as yet unproven' (OX Consultants 2004: 20). This was reinforced later that year by a recommendation from the communications regulatory body, ComReg, which suggested that the spectrum allocated to DAB should be used in fact for a terrestrial digital *television* system, DVB – which can, of course, also deliver radio stations alongside its digital TV channels (ComReg 2004). Only with the potential reallocation of unused frequencies at the ITU's 2006 Regional Radio Conference (RRC06) was interest sharpened in advancing digital broadcasting in Ireland (RTÉ 2005: 14). New broadcasting legislation in 2007 provided for the introduction of digital terrestrial television and sound broadcasting. A second phase of digital radio deployment was initiated when RTÉ re-launched its DAB service in 2007, initially as a trial service, this time on two multiplexes with a total of eleven stations, including a mix of public and commercial stations. Following its conclusion in November 2008, RTÉ proceeded on its own with a DAB service in the three main urban centres of population, without the involvement of commercial radio. Additional incentives for the licensing of commercial stations on DAB are contained in new legislation with a promised six-year extension for analogue licence-holders who take up the option. To date, however, no applications for commercial involvement in DAB have been received.

In Denmark and Finland, the FM spectrum also has limited capacity for further expansion. In Denmark commercial radio, which operates predominantly in local coverage areas, was effectively ruled out of DAB's early development, as the platform's variable efficiency in its use of frequency encourages multiplexes to be arranged to cover either the whole country or at least large regions. The DAB multiplexes are thus populated mainly by existing and new public service channels from DR, with the exception of provision for the two national commercial stations, Radio Nova (SBS) with one channel, and Radio 100FM with two channels. Other commercial stations, covering small geographical areas, will not find space on the DAB platform until the introduction of new L-band frequencies towards the end of the decade.[3] With commercial radio's audience share at around 30%, and the 2005 demise of Sky Radio, the demise of the successor TV 2 Radio after only one year in service and with the third operator in five years Radio Nova, DR remains still the driver for DAB in line with Danish digital broadcasting policy (Jauert 2008). The DAB structure in Finland posed similar difficulties for local commercial stations: the geographical coverage area of the DAB multiplexes were regional in nature rather than local, vastly exceeding the existing reach of local stations. In 1995, the Finnish Ministry of Transport and Communications stated that 'either the Finnish commercial local radio stations must be turned into regional stations with significantly larger coverage areas, or they need another, alternative frequency allocation' (MINTC 1995: 24).[4] As in Denmark, local stations in Finland would have to wait for L-band frequencies to become viable in future years. With the two largest commercial radio companies withdrawing plans for DAB, it was again left to public broadcaster YLE to develop the DAB networks carrying its national and regional stations (Ala-Fossi 2001: 12).

This mismatch between DAB's 'local' coverage and existing local analogue stations clearly makes DAB unattractive to these countries' commercial radio stations. Even in the UK, where an allocation of five multiplex frequencies for local services, in addition to the two national frequencies, allowed greater flexibility in planning coverage than in the other countries (where each had a total allocation of two frequencies), and the geographical reach of DAB transmissions has tended to match the bigger 'local' or regional stations, leaving out the smaller commercial and the new community stations. Augmenting this more suitable multiplex geography, the United Kingdom had additional policy strategies to encourage commercial radio to become a major part of DAB from the outset. In addition to straightforward commercial incentives (awarding of multiplex licences on a commercial basis; automatic renewal of analogue radio licences), the digital radio regulator, the Radio Authority, was to take a 'lighter touch' to regulation in comparison with its role in analogue radio. For example, decisions about which stations should be carried on a particular multiplex were largely to be determined by the commercial multiplex operators, and such decisions were to be made on a commercial basis (with the sole exception of the obligation to carry the local BBC station if it covered the same locality). The regulator was unable to intervene in such decisions:

> ... the Authority is not empowered to specify the types or numbers of digital sound programme or additional services which it expects to be provided on a multiplex ... decisions about the choice and nature of sound programme and additional service providers are for the multiplex licence applicant to make. (Radio Authority 2001: 21)

When Ofcom took over the Radio Authority's role in 2003, it also argued that there was less need to regulate commercial radio:

> The general principle ... is that as spectrum constraints lessen, the need for regulation decreases, as the market provides ever wider choice. It could be argued that, as digital take-up grows, the need for regulation on analogue platforms will decrease, as listeners can experience the wider choice available on all platforms. (Ofcom 2004: 57)

The lack of regulation on the composition of the digital radio multiplexes has allowed the large commercial radio groups which own the multiplex licences to increase networking between stations, and to turn local analogue stations into 'quasi-national' stations on digital radio. Thus, stations which broadcast in analogue only in London can be heard across the country on DAB multiplexes, becoming more valuable to their owners in attracting advertising revenue. Similarly, new digital-only stations launched by the radio groups are carried on most of those same groups' multiplexes in different parts of the country. While this makes obvious commercial sense, it compounds the exclusion from DAB of smaller commercial and community stations already facing problems with multiplex coverage areas. This segregation has been noted by the Digital Radio Working Group set up by the UK government, which recommended that smaller stations should remain on analogue FM for the foreseeable future (DRWG 2008). Hence, for local and community radio stations, DAB presents some difficulties rather than opportunities, such as being 'left behind' on what might be regarded as an outdated technology.

Among the countries in Mid- and Southern Europe, Germany shows some structural resemblances with Spain, but the role of the governments, the regulatory bodies and the main players in the media landscape have been quite different. In Germany, the industry, the commercial and the public radio sector have acknowledged that DAB has stalled, and therefore DAB+ should be the main standard for digital radio in Germany. But this does not mean that broadcasters in some regions will not use other technologies such as DRM, DRM+ and DVB-T, offering more digital radio services and options to German consumers.

The Spanish situation is special compared to other countries in Europe having selected DAB as the main standard. In early 2008, the World DAB Forum assessed the DAB situation in Spain to be 'at a standstill, as the major radio groups remain unsure due to the perceived threat to their markets' (World DAB 2008). The historical background for this is that radio in Spain, since the founding years in the 1930s, has been stuck halfway between the North American commercial model and the European public service model (Bonet et al. 2009:11), and has been an example of coexistence ever since. Due to the large quantity of local stations, the Spanish radio market appears atomized, based on chain broadcasting with co-existing nationwide public and private broadcasters, all with a strong dependency on advertising. Since digitalization was initiated as a top down process from the national government, the commercial market reacted like the US-market did, or probably would have done in a similar situation. In fear of losing its foundation in the well-established analogue market, commercial broadcasters, especially, engaged only half-heartedly in promoting DAB. Whereas US-broadcasters, supported by the FCC, promoted the IBOC standard that combined the analogue and digital frequency spectrum and thus kept the market basis intact, Spanish broadcasters were forced into the dominant European standard. In Spain, therefore, the introduction of DAB happened in a way that was 'counter to the market'. Bonet et al. enumerate the following issues as crucial for the failure of DAB in Spain:

- Total lack of coordination among all 'the players'
- Poor coverage and reception: DAB works well for national services, but is not the best for local radio
- No social need for DAB, no new services compared to FM were offered – no new and different scheduling and no new programme content (Bonet et al. 2009: 24)

While similar conditions existed in other countries in Europe, the totality of strategic errors, its top-down process, as well as the special composition of the Spanish radio market, has led to this *de facto* collapse of DAB in Spain without any obvious remedy. By contrast, other countries in almost the same situation – France, Germany and Sweden – are all trying to introduce new or supplementary digital standards in order to reinvigorate the digitalization of radio

Conclusions

The four countries in the centre of this comparative study reveal quite different approaches to the launch of digital radio. Fifteen years ago, each was beginning from more-or-less the same position: the DAB system was proven technically but not commercially, and was to be introduced into an uncertain market. Now four different outcomes have been reached, with two countries leading by far, while a third remains at the starting gate and the fourth has given up DAB in favour of DVB-H (and DVB-T). The differences result from the varied policy decisions taken by the governments and broadcasters in those countries, policies which both reflected and also reinforced differences in the existing organisation and structures of radio in each country, but which were also constrained by the technical limitations of the DAB system. The UK government has been most explicit in its policy on digital radio. In order to secure the growth of digital radio, it has been organised in such a way that commercial radio companies would be encouraged to participate, a position in evidence also in the organisation of digital television (Galperin 2004). In particular, the size of the commercial radio industry and its relatively high concentration of station ownership meant that, as UK DAB policy evolved, the larger commercial radio companies were able to expand and strengthen their position in relation to the BBC in a deregulating environment (Lax 2009).

In contrast, development of DAB in Denmark has been driven by the public service broadcaster DR, with little support from commercial radio. While DR maintains a substantial share of analogue listening – the highest of the four countries considered here – the arrival of local commercial radio, and still more recently of national commercial radio competing directly with its own national and regional stations, has forced DR to respond and introduce radical changes (Jauert 2003). It has expanded substantially its provision on DAB and other digital platforms such as Internet radio since the relatively late launch in 2002. Here, the relative indifference of commercial radio to DAB has not inhibited growth; instead it is this augmentation of an already popular public broadcaster's output that appears to have steadily driven recent sales of receivers to the point where approximately 30% of households have a DAB receiver, and thus exceeding the United Kingdom, with a penetration of 20% (EBU/UER 2007).

In Finland, which has a similar radio landscape to Denmark, digital radio has largely been underpinned by a policy commitment to increasing the availability and take-up of digital TV, which of course can also deliver digital radio stations. While public broadcaster YLE did begin transmitting services on DAB from 1998, at the same time a report for the Finnish Agency for Technology and Innovation emphasised the importance of digital TV for the delivery of multimedia services, and by 2001 YLE had decided against further expansion of the DAB network (Guy and Stroyan 1998; Autio 2001). As we have noted, commercial radio did not invest in DAB and in 2003 the first commercial digital radio licences were awarded instead for delivery over the digital terrestrial television (DVB-T) platform rather than DAB. In 2004, the Finnish government's communications ministry concluded that, 'at this stage there are no particular reasons to hasten the digitalisation of radio' (MINTC 2004: 12). In addition, Nokia, Finland's largest company and a world leader in telecommunications, had already taken an early decision in 1997 to withdraw from development of DAB devices in favour of concentrating on the DVB-T television platform with its greater potential for multimedia services. Its subsequent development of a mobile version of DVB-T, DVB-H (digital video broadcasting, handheld) for delivery of multimedia (including radio) has reinforced YLE's 2005 decision to end its DAB transmissions, effectively signalling the end of DAB in Finland, with the regulator also indicating tacit support for the DVB-H alternative (MINTC 2003; Ficora 2004). In this context it is easy to understand why the two largest commercial radio companies in Finland were eager to start digital radio broadcasting on the new DVB-H network in 2006.

The slow pace of development of DAB in Ireland perhaps most clearly demonstrates the combination of factors influencing digital radio policy. With FM spectrum still available, and no interest from commercial radio in DAB because of its unsuitability for local broadcasting, the incentive for the development of DAB was more a rather abstract case of not being 'left behind' in the technological onward march rather than a clear economic argument for DAB's potential for expansion. The advice received in government consultations was generally cautious with some conflicting recommendations: initially, to wait and observe how DAB fared elsewhere; later for either a steady roll out of DAB or, conversely, for the use of the spectrum for digital TV. The industry in general has regarded the substantial costs of developing DAB as unjustified. Government policy has remained largely reticent on the matter.

By including a number of other countries in this comparative study we have tried to add some wider perspectives to the very complex digital radio landscape in Europe. Added to the differences we have traced emerging from specific media cultures; the size of the media markets; legislation; and media policies and politics; an overall assumption of the causes of the relative success, and not least the many failures and setbacks for DAB, could be 'that the technology has been seen as an autonomous power, whereas some basics of human behaviour have been ignored by some of the actors for too long' (Kleinsteuber 2006:144).

We have presented here an account of a situation still under development. While the radio industry is continually changing in each of these four countries, the technological possibilities are changing faster still – during the course of this research, a number of new digital radio standards have been developed around the world. However, the experiences of the introduction of DAB, the 'original' and certainly most long-lived standard, highlight some of the complexities of the paths along which any new standards must evolve.

Notes

1. This chapter is a revised and updated version of Lax, S., M. Ala-Fossi, et al. (2008) previously published in the journal *Media Culture and Society*.
2. Deutschlandradio is a national German public radio broadcaster. It operates two national networks, Deutschlandfunk and Deutschlandradio Kultur. It is a merged company of former Eastern and Western German boadcasters, established after the reunification in 1994.
3. In the higher-frequency L-band, DAB can in principle operate over smaller coverage areas, although inefficiently. However, the L-band has hitherto not been used extensively for broadcasting and so it is unclear how appropriate it may prove to be in this role.
4. The problems with DAB in small-scale broadcasting had been pointed out already in 1992 in a report to CDMM, the Steering Committee on the Mass Media of the Council of Europe (Gronow, Lannegren & Maren 1992).

References

Ala-Fossi, M. (2001), 'From channel competition to cluster competition: Finnish and American radio formatting 1985–2000', Presented at the *46th Annual Convention of the Broadcast Education Association*, 20–23 April, Las Vegas.

Ala-Fossi, M. and P. Jauert (2006), 'Nordic Radio in the Digital Era', in U. Carlsson, *Radio, TV & Internet in the Nordic Countries Meeting the Challenges of New Media Technology*, Göteborg: Nordicom, pp. 65–87.

Autio, M. (2001), 'Digiradio soi toistaiseksi harakoille/For the time being, digital radio broadcasts for nobody', *Helsingin Sanomat*, 11 November.

Bonet, M., M. Carominas, I.F. Alonso and M. Dìez (2009), 'Keys to the Failure of DAB in Spain', *Journal of Radio and Audio Media*, 16: 1, pp. 83–101.

Carter, M. (1999), 'The changing face of radio', *Director*, 42:6, March 1999.

ComReg (2004), *ComReg Response to Consultation on Frequency Spectrum Policy for Digital Broadcasting* [Document 04/93], Dublin: Commission for Communications Regulation.

CRAOL (2004), *Response to the review of radio licensing report*, Dublin: CRAOL.

Crisell, A. (2002), *An Introductory History of British Broadcasting; Second Edition*, London: Routledge.

DRWG (2008), *Digital Radio Working Group Final Report*, London: Department of Culture, Media and Sport.

EBU-UER (2007), *Public Radio in Europe 2007*, Geneva: European Broadcasting Union.

Ficora (2004), 'Tukholman tv-taajuussopimuksen uudistamista valmistelevan työryhmän (ST61REV) 3. kokous, kokousmuistio ST61REV 02(04)/Minutes of the Third Meeting of a Preparatory Working Group for the Renewal of the Stockholm Television Frequency Agreement. ST61REV 02(04)', Helsinki: Finnish Communications Regulatory Authority.

Hedges, Michael (2009), 'Digital radio for toi et moi', *Followthemedia*, http://www.followthemedia.com/alldigital/DRFrance29052009.htm. Accessed May 29, 2009.

Galperin, H. (2004), *New Television, Old Politics: The Transition to Digital TV in the United States and Britain*, Cambridge: Cambridge University Press.

Guy, K. and J. Stroyan (1998), 'Digital Media in Finland', *Evaluation Report Technology Programme Report 12/98*, Helsinki: TEKES (Finnish Funding Agency for Technology and Innovation).

Gronow, P., G. Lannegren and L. Maren (1992), 'New Technical Developments in the Sound Broadcasting Sector and their Impact on Mass Media Policy', Council of Europe, CDMM, Strasbourg, 22 September 1992.

Jauert, P. (2003), 'Policy development in Danish radio broadcasting 1980–2002. Layers, Scenarios and the Public Service Remit', in Gregory Ferrell Lowe and Taisto Hujanen (eds), *Broadcasting and Convergence: New Articulations of the Public Service Remit*, Göteborg: Nordicom, pp. 187–203.

Jauert, P. (2008), 'Fra broadcast til podcast – digital radio i Danmark', in F. Mortensen (red.), *Public service i netværkssamfundet*, 'From Broadcast to Podcast – Digital Radio in Denmark', in F. Mortensen (ed.), *Public Service in The Network Society*, Frederiksberg: Forlaget Samfundslitteratur, pp. 103–148.

Kleinsteuber, H.J. (2006), 'A Great Future? Digital Radio in Europe', in Frédéric Antoine (ed.), *Nouvelles Voies de la Radio/The Way Ahead for Radio Research, Recherches en Communication*, 26, Louvain-la-Neuve, 2006, pp. 135–144.

Lassila, Anni (2008), 'Kännykkä-TV hake edelleen yleisöä ja toimintamallia/Mobile phone TV still in its infancy', *Helsingin Sanomat*, 24 November, http://www.hs.fi/english/article/Mobile+phone+TV+still+in+its+infancy/1135241357566. Accessed 29 May 2009.

Lax, S., M. Ala-Fossi, P. Jauert and H. Shaw (2008), 'DAB: the future of radio? The development of digital radio in four European countries', *Media Culture Society*, 30: 2, pp. 151–166.

Lax, S. (2009), 'Digital radio and the diminution of the public sphere', in R. Butsch (ed.), *Media and the Public Sphere*, Basingstoke: Palgrave (Forthcoming).

MINTC (1995), *Yleisradiotoiminnan strategiaselvitys. Radio ja televisio 2010 / A Strategy Report on Broadcasting. Radio and Television 2010*, Helsinki: Ministry of Transport and Communications.

MINTC (2003), *A Fourth Digital Broadcast Network: Creating a Market for Mobile Media Services in Finland*, Helsinki: Ministry of Transport and Communications.

MINTC (2004), *Radiotoiminta 2007: Työryhmän ehdotukset / Radio Broadcasting in 2007: Proposals of the Working Group*, Helsinki: Ministry of Transport and Communications.

Ofcom (2004), *Radio – Preparing for the Future. Phase 1: Developing a New Framework*, London: Office of Communications.

OX Consultants (2004), *Final Report: Review of Licensing of Radio Services in Ireland. A Study for the Department of Communication, Marine and Natural Resources*, Dublin: Department of Communications, Marine, and Natural Resources.

RAB (2009), 'Commercial Radio Revenues', www.rab.co.uk. Accessed 30 April 2009.

Radio Authority (2001), *Local Digital Radio Multiplex Service Licences. Notes of Guidance for Applicants*, London: Radio Authority.

RAJAR (2009), 'Quarterly Summary of Radio Listening. Survey Period Ending 14 Dec. 2008', www.rajar.co.uk. Accessed 30 April 2009.

RTÉ (2005), *Digital Television and Radio Services in Ireland: an Introduction*, Dublin: RTÉ.

SOU (2002), 'Digital Radio. Kartläggning och analys. Delbetänkande av Digitalradioutredningen / Digital Radio. Survey and analysis. Interim Report from the Commission on Digital Radio', Stockholm: The Swedish Ministry of Culture, http://www.regeringen.se/sb/d/108/a/1482;jsessionid=aFcBKuipt4x4.

The Swedish Radio and TV Authority (2008), *The Future of Radio*, Stockholm: The Swedish Radio and TV Authority.

World DAB (2005), 'New wave of DAB legislation and developments worldwide', Press release, 14 April 2005.

YLE (2004), 'Digitaaliradio. Palvelut / Digital Radio. The Services', www.yle.fi/digiradio/palvelut.htm. Accessed 27 May 2005.

Chapter 6

Digital Radio Strategies in the United States: A Tale of Two Systems

Alan G. Stavitsky & Michael W. Huntsberger

The case of digital radio in North America illuminates the contradiction between federal communication policy ideals and *realpolitik*. The policy of the United States government gives official imprimatur to robust competition and to local broadcasting that serves 'the public interest, convenience or necessity,' in the words of the federal licensing standard (Radio Act of 1927). In the decades since the passage of the Federal Radio Act, notions of capitalism and communication have intertwined as they have been set down in the re-conceptions and revisions of the original statute. Within the local marketplace, the theory goes that unfettered capitalism will lead to efficient exchange of goods and services, while free and open discourse will yield the best ideas to promote the democratic process (Stavitsky 1994). On the foundation of this theoretical model, broadcast stations in the United States have always been licensed at the level of the local community. To ensure competition, regulators have historically set and enforced limits on the number of stations any individual or agency could own.

Broadcast regulation in the United States, however, has long been marked by tension between the ideals of localism and competition, and the lure of centralization. While national broadcasting systems dominated the development of European radio, U.S. radio began in the early 1920s with independent local stations drawing upon local voices. The rhetoric of the time reflected utopian notions of radio as a conduit of civic discourse through which citizens would deliberate the public affairs of the community. In practice, however, network broadcasting developed rapidly and the 'chains', as the first national broadcasting corporations were originally known, became the dominant source of programming within radio's formative first decade. In addition to this centralization of content, control of stations became increasingly concentrated as broadcasters successfully lobbied for gradual relaxation of ownership limits. For radio, this deregulatory trend culminated in passage of the Telecommunications Act of 1996, which eliminated all restrictions on national ownership, while retaining limits within any particular market (Telecommunications Act of 1996).

This capsule history points to the ambiguity of U.S. communication policy. The focus of regulation remains on local service, while content and control are largely centralized. Though the metaphor of the 'marketplace of ideas' implies robust competition, ownership has been allowed to concentrate. Further, the Federal Communications Commission, while seeking to bring the benefits of new technology to industry and public, has historically been reluctant to set technological standards, preferring to let the market decide. At times this lack of symmetry between regulators and industry has had significant consequences. The commission's failure to set standards for AM stereo systems in the 1980s helped doom the technology to irrelevance (Sterling and Kittross 2002: 570). Similar ambiguities have characterized the emergence of digital radio, as U.S. regulators have been unprepared or unwilling to face the challenges presented by new technologies, and have consistently deferred to industrial imperatives as they established a policy framework.

Stumbling toward a digital radio standard

Initiatives to migrate broadcast radio in the United States from analogue to digital systems began in the early 1980s, concurrent with similar efforts in Europe and Asia. With the introduction of audio compact disc players to the consumer market in 1982, U.S. broadcasters sought methods to upgrade the audio quality of their services in order to provide CD-quality sound (Radio World 2008). Broadcast transmission and reception systems of the time were inadequate for the task, leading to a variety of experimental approaches to 'going digital'. To accommodate the additional bandwidth necessary for digital encoding and transmission, Boston public radio station WGBH experimented with modulating a digitized audio programme stream on the licensee's UHF public television channel. While the experiment was considered successful, the broadcasts were available to an audience of perhaps a few hundred people who owned professional digital audio processors. WGBH was also among the first to use digital systems to distribute programme material between remote and head-end facilities (Bunce 1986: 21).

During this period of digital experimentation, the historic regulatory structures and policies that had governed U.S. broadcasting since the 1930s were undergoing fundamental and profound change. With the inauguration of President Ronald Reagan and the subsequent appointment of Mark Fowler as chair of the Federal Communications Commission in 1981, broadcasting licenses that had been rigorously regulated by the F.C.C. became commodities that could be easily traded on the open market. Widely credited for referring to television as 'a toaster with pictures', Fowler de-emphasized the Commission's policy research and recommendation functions in favour of market-based solutions derived from the practices of industry (Boyer 1987: C15). Notable for decisions that eliminated requirements for public service programming, lowered standards for license renewal and removed restrictions on the sale of licenses, the F.C.C. under Fowler nurtured an environment that encouraged private interests to take the lead in the development of digital broadcasting in the United States.

The challenge emerged in 1990, as the F.C.C. considered whether to authorize both terrestrial and satellite-delivered digital radio services. Somewhat surprisingly, the first initiative to come before the Commission did not originate with one of the major U.S. broadcasting companies or equipment manufacturers, but from a start-up: Satellite CD Radio Inc. petitioned the F.C.C. to allocate space in the S-band between 12.2 and 12.7 GHz for the transmission of signals to geostationary earth-orbiting satellites capable of transmitting a nationwide, multi-channel digital audio service directly to consumers. The Satellite Digital Audio Radio System [S-DARs] – satellite radio – would be available throughout the continental United States, allowing the consumer to travel coast to coast without experiencing interference or service interruptions (Huntsberger 2001). Charging that the Satellite CD plan threatened the local service of 12,000 terrestrial radio broadcasters, the National Association of Broadcasters asked to F.C.C. to dismiss the application. At the time, the president of Satellite CD observed that the commercial broadcasters were not motivated by aspirations to preserve the traditions of local service or the public interest, but rather by the more essential desire to restrict competition (New York Times 1990: D4).

On behalf of its members, the vast majority of whom are commercial enterprises, the N.A.B. considered a variety of systems that could be capable of providing CD-quality digital audio broadcast services, including delivery by cable, satellite, and terrestrial channels. As the World Administrative Radio Conference moved toward adoption of the DAB system, the N.A.B. was pushing for the adoption of the Eureka 147 system as the U.S. standard (New York Times 1991: 115). But L-band DAB faced a host of challenges. Because Eureka 147 was suitable for terrestrial and satellite transmission, existing broadcasters feared that the technology might provide parity for S-DARS. In addition, broadcasters viewed the multiplex capability of Eureka 147 as an opportunity for new terrestrial competition, and worried that Eureka 147 allocations might not match existing coverage.[1] Public agencies had other concerns. Following the success of Operation Desert Storm, a coalition of forces in the administration of George H.W. Bush moved to protect the L-band for the use by the U.S. Department of Defense for 'aeronautical flight-test telemetry' for the development of new, advanced weapons systems (Belsie 1992: 9).

As the debate over spectrum allocation and transmission standards proceeded, an initiative developed jointly by commercial ownership groups CBS Radio, the Gannett Company and Group W Broadcasting proposed a system that would allow for simultaneous transmission of analogue and digital signals on existing FM frequencies, and possibly on AM channels as well. Such an in-band on-channel [IBOC] system offered the possibility that the move from analogue to digital technology could be accomplished without any dislocation to the existing marketplace for broadcast radio programming and advertising. Specifically, the IBOC system promised existing licensees the chance to enhance the sound of their broadcasts and carry additional services on their signals without opening channels for new competitors. But it would not be easy: one CBS executive compared the challenge to 'fishing out millions of needles every second from an endless line of haystacks' (Andrews 1992: D8).

Throughout the 1990s, the N.A.B. considered several approaches to IBOC digital radio. But the primary policy agenda for commercial radio broadcasters was characterized by their united and vocal opposition to satellite radio. At every turn, the N.A.B. asserted the position that nationwide satellite services would undermine the ability of local broadcasters to 'attract listeners, sell advertising and maintain their viability' (Andrews 1992: D1). While engineering trials for satellite broadcasting proceeded smoothly, the progress of satellite radio in the sphere of public policy slowed to an almost glacial pace. The F.C.C. asserted that 'existing radio broadcasters can and should have the opportunity to take advantage of new digital radio technologies', and acknowledged industry concerns that the national footprint of satellite radio posed a competitive threat to local stations, reaffirming localism as a 'touchstone value' (F.C.C. 1999: 4). At the same time, while recognizing the terrestrial and satellite services 'would compete to some extent', the Commission concluded that satellite radio would complement broadcast radio 'by providing regional and national services', and that IBOC DAB was not yet technically feasible. Backed by arguments for technology advancement and market diversity, the proponents of satellite radio ultimately succeeded, convincing the F.C.C. to allocate spectrum and consider license applications from four private companies (Andrews 1995: D14). Two of these companies survived as XM Radio, launched in November 2001, and Sirius Satellite Radio, launched in 2002 – the dramatic launches garnering considerable attention from the press and investors on Wall Street. Though they were separate and competing companies, the fates of the two digital services would remain connected.

Faced with the realities of competition, the N.A.B. moved ahead with efforts to develop a workable IBOC system. It would be nearly four years until the CBS-Gannett partnership, operating as USA Digital Radio, filed the first Petition for Rulemaking with the F.C.C. to permit IBOC as the terrestrial digital radio broadcast standard in the U.S (Desposito 1999: 45), and another three years before the Commission approved an IBOC system for use in the United States (Feder 2002: C3). While IBOC on the FM band received the approval of the National Radio Systems Committee, the influential engineering group could not endorse similar technology for AM broadcasting, citing night-time interference problems. Because one of the selling points of IBOC had been improved fidelity for AM broadcasters on a par with FM-band signals, any strategy that might move forward with FM-IBOC only was questionable. Nevertheless, IBOC was presented to broadcasters with great fanfare at the 2002 NAB convention as a fully operational system and approved by the F.C.C. later that year, despite AM transmission problems and a lack of testing with consumer receivers.

Significantly, the Commission's decision did not include a calendar mandate for digital conversion. With a tip of the hat to former Chairman Fowler's bedrock belief in the power of free markets, regulators followed the preferences of industry leaders and turned to the forces of supply and demand to catalyze the adoption of digital radio in the U.S. The approved IBOC system would be manufactured and distributed exclusively by iBiquity Digital, and available only under license from the manufacturer. The rush to market was driven by a desire to reassure Wall Street of the continued viability of terrestrial radio in the digital age, even as satellite radio systems and services began to appear in the consumer market (Ala-Fossi and Stavitsky: 2003).

As commercial radio interests took more than a decade to roll out their chosen digital system, a variety of chipmakers and equipment manufactures capitalized on the slow, deliberate pace of IBOC development. Motorola, Blaupunkt and Texas Instruments were among the companies that developed technologies for digitizing, filtering and manipulating analogue radio signals to improve reception and audio quality (Feder 2002: C2). The public release of the World Wide Web allowed stations to deliver digitally-encoded audio programming directly to personal computers, and created a global delivery platform for audio content and services. Software designers developed algorithms that reduced the size of digital audio data files so they could be economically captured, manipulated, stored and retrieved by personal computers, without apparent loss of sound quality. By 2001, when Apple released the iPod portable player and the iTunes software that allowed consumers to easily select, purchase, store, and replay thousands of digitally encoded songs on demand, the consumer market for audio media had entirely outstripped the scenarios envisioned by U.S. commercial broadcasters, and the N.A.B. Satellite radio had become just one of a host of digital audio technologies and services competing with terrestrial radio for the attention of American consumers.

The symbiosis of the terrestrial and satellite digital systems re-emerged with the 2008 merger of XM and Sirius into a single, corporate entity. When the F.C.C. issued the original licenses to the two companies, it did so with the stipulation that one company would not be permitted to acquire control of the other. This stipulation was intended to ensure competition and mollify terrestrial broadcasters, but once the business plans of the two companies were set into motion, both satellite providers saw their operating costs soar like their respective launch vehicles, and eventually Sirius and XM sought government approval to merge in 2006. This led to eighteen months of official deliberation by the U.S. Justice Department which considers anti-trust matters. At issue was the definition of market; were XM and Sirius in effect competing with local radio stations, transmitting from 25,000 miles up? Viewed in this way, a satellite radio monopoly could create a formidable challenge that threatened the economic viability of local terrestrial broadcasters and, by extension, the bedrock value of localism.

But Sirius and XM argued that satellite radio is simply another aural alternative in an audio marketplace that includes not just terrestrial radio, but also Internet audio services, iPods and other digital players, and audio-enabled cell phones. The Justice Department, and eventually the F.C.C., took the latter view, ruling that the merger would not hurt competition. 'In several important segments of their business, with or without the merger, the parties simply do not compete today,' said Thomas O. Barnett, head of the Justice Department's anti-trust division. 'Some people may view iPods as a particularly good alternative. They may view HD radio as an alternative' (quoted in Philip Shenon, 3/25/08, NYT, Justice Dept Approves XM Merger with Sirius). As Sirius and XM move to combine their operations, sales, marketing and customer service functions, the short-term benefits of the merger will accrue first to the business enterprises. For consumers, the full value of a combined service and its competitive impact on the market, will not be apparent until receivers capable of picking up both satellite services become available, perhaps in the next year (Pizzi 2008: 42).

Playing catch-up

Given the long road to IBOC implementation, U.S. radio broadcasters have moved relatively quickly to bring the technology to market and stimulate consumer interest. Capitalizing on burgeoning interest in the federally mandated conversion to High Definition Television [HDTV], iBiquity rechristened the IBOC system as HD Radio.[2] By October 2003, 280 stations in more than 100 markets had purchased licenses for HD Radio technology, and but only 70 were broadcasting digitally encoded FM signals (Berger 2003: G3; iBiquity 2003). HD Radio-capable home receivers appeared at the 2004 Consumer Electronics Show, and several automobile manufacturers carried HD Radio receivers in their 2004 models. But consumer interest was marginal at best. The receivers were expensive: Kenwood's KTC-HR100 add-on tuner for car stereos carried a price tag of $350 (Berger), far more than add-on units for satellite radio, and a substantial price to pay for CD-quality audio in the listening environment of a moving car with an 80+ dB noise floor. HD receivers designed for home use entered the market with retail prices of $500 or more (Fleishman 2005: C11). More recently, prices for home and auto receivers have moderated, with some units now costing less than $100 (iBiquity 2009). However, power requirements for HD chips exceed specifications appropriate for personal devices, and no manufacturer currently offers a portable HD unit that can complement or compete with Apple's iPod and similar digital players, shutting HD broadcasters out of this most significant market.

Faced with such challenges, in 2005 some of the largest commercial radio companies in the U.S., including Clear Channel Radio, Entercom and Infinity Broadcasting, formed the HD Digital Radio Alliance, a co-ordinated, national strategic marketing campaign to 'accelerate the rollout of HD Digital Radio' (HD Digital Radio Alliance 2005). Paradoxically, the campaign was undertaken at a political level in the name of protecting the U.S. ideal of locally focused and controlled radio stations, despite criticism that the consortium violates anti-trust law. The strategy of the Alliance positioned HD in opposition to satellite radio, offering a parallel benefit – programme diversity – without the cost of a satellite radio subscription. Beginning with radio spot advertising on 280 stations in 28 markets, valued at $200 million, the campaign touts the benefits of HD to listeners, and promoted the availability of receivers for new and existing cars and homes (HD Alliance 2006). The effort provides broadcasters with logos, brochures, print ads, web banners and other graphic materials, an audio podcast, a video with tips for consumers, and links to rebate offers from HD radio manufacturers and dealers (HD Alliance 2009).

While the Alliance touts the efficacy of these efforts, a 2007 study by independent research firm Bridge Ratings concludes that market penetration by HD radio lags far behind competing technologies, especially analogue AM and FM radio, which continues to serve over 90% of Americans each week. iPods and other digital players reach 30% of the population, and satellite radio penetrates almost 5%. In comparison, HD Radio reaches less than 1% of Americans: Bridge estimates that 450,000 Americans listen to HD radio every week, compared to 57 million that listen to some form of Internet radio (Bridge Ratings 2007). One analyst asserts that the steady, deliberate promotional strategy of the Alliance is 'difficult to support'. Such a long-term, continuous campaign is perhaps the hardest type of promotional exercise to undertake successfully, as opposed to the sort of short-term, high-visibility strategies that are preferred by marketing professionals (Pizzi 2008: 15). These developments suggest that HD radio has a long way to go to catch up with other digital audio technologies available in the U.S.

U.S. public radio broadcasters have focused primarily on the feasibility of using the secondary audio channels [SACS] on HD Radio signals to provide additional programme services to niche audiences. Throughout 2003, field tests conducted by NPR Labs examined the performance of HD channels carrying two programme streams. The findings of the Tomorrow Radio Project demonstrated that spectrum split into two streams was sufficiently robust to provide high-quality reception in mobile environments (NPR Labs 2004). In subsequent years, NPR Labs has become a leader in HD Radio research, publishing findings on coverage and interference.

Digitization also offers the promise of new public-service applications. Many public radio stations use their analogue sidebands to broadcast programmes for visually impaired listeners, but these 'radio reading services' require distribution of specially equipped receivers. HD Radio accommodates these services on multicast side channels (though this does by necessity still require listeners to obtain a digital receiver). Further, NPR Labs, the research-and-development wing of National Public Radio, collaborated with Harris Corporation, a U.S. communication equipment manufacturer, and Towson University engineers to develop an audio-captioning system that does, in effect, subtitle radio programming for hearing-impaired people (CITE). The system was demonstrated in live coverage of the U.S. presidential election returns in November 2008 ('Captioned radio broadcast to enable millions of deaf and hard-of-hearing to experience NPR's live coverage of presidential election for the first time,' 21 October 2008, news release, International Center for Accessible Radio Technology, http://I-cart.net).

International concerns

Presently, no nation or governing body beyond the United States has considered HD Radio as a standard for digital broadcasting. Yet the nation's position as a dominant economic and political power, and one of the world's largest markets for broadcasting technology for consumers and professionals, presents considerable challenges and opportunities for other digital radio interests around the world. In addition, the U.S. shares two borders; one with Canada totaling almost 9,000 kilometres, and one with Mexico of just over 3,000 kilometres. This shared geography poses a number of issues related to spectrum allocation, cross-border interference, and the availability of services and equipment in professional and consumer markets.

For many years, Canada embraced the Eureka 147 DAB standard for the transition to digital radio broadcasting. But a 2006 report on the future of broadcasting in Canada recognized that analogue FM remained a dominant form of radio, and that Canadian broadcasters have 'adopted new technology platforms through Internet streaming and podcasting, as well as entering into content partnership arrangements with other undertakings including satellite providers', particularly those based in the U.S. (O'Neill 2008: 32). Recognizing the new realities in the digital audio marketplace, the Canadian Radio-Television and Telecommunications Commission revised the nation's digital radio policy to consider DAB as one of a number of appropriate technologies, including HD Radio that might be adopted in Canada. The CRTC also announced it was prepared to authorize IBOC services in Canada (Radio 2006). Field trials conducted in Toronto in 2007 concluded that the technology poses interference problems for existing analogue FM services, and that implementation will have to be accompanied by review and revision of spectrum management rules (Bouchard 2007). Mirroring the U.S. experience, the Bouchard report observes that consumers will ultimately decide whether HD services will be viable in Canada. To date, no agency has applied for authority to test or adopt HD Radio for any location in Canada.

More recently, the Federal Telecommunications Commission of Mexico authorized HD Radio transmission within 320 miles of the U.S. border. Recognizing 'the extent of the development and implementation of the IBOC system' in the U.S., COFETEL elected to forego further research and field trials in advance, taking 'decisive action' to bring HD Radio to listeners in northern Mexico (Radio 2008). Mexican regulators directed stations interested in HD transmission to request authorization and assist with assessments of the technology. Conveniently for American broadcasters, the 320 km limit accommodates the signals of stations broadcasting from San Diego, Tucson, El Paso and other metropolitan areas along the border.

Social implications: Radio's 'third chance'?

The British broadcaster Charles A. Siepman, who was called upon by the F.C.C. to write the well-regarded (but ultimately ignored) *Blue Book* on public service in American broadcasting, referred to the development of FM broadcasting in the 1940s as 'radio's second chance'. He wrote of the social potential of FM to redeem radio's promise after the AM band became awash in mass entertainment and advertising (Siepman 1946). If FM represented radio's second chance, might the digital transition constitute the medium's third chance?

It's clear that digital radio offers some of the same advantages noted by Siepman in the case of FM – additional channels of communication through multicasting, and improved audio quality. Satellite providers tout the availability of scores of channels, more than 170 on XM and more than 130 on Sirius (some channels were heard on both services after the merger). 'Everything worth listening to is now on Sirius,' reads the corporate slogan (www.sirius.com). Talk programming crosses the political spectrum, from the Sirius Left channel of liberal content, to Sirius Patriot, a conservative outlet. Like oldies music? XM offers separate channels for each decade from the 1940s through the 1990s, as well as broadcasts of the Super Bowl in 10 languages, including Flemish, Hungarian and Mandarin Chinese (www.xmradio.com).

HD Radio, while offering fewer channels because of technical limitations, nonetheless seeks to provide programme alternatives beyond the usual broadcast fare. In Portland,

Oregon, for example, the 23rd-largest radio market in the U.S., fourteen broadcasters were multicasting secondary channels at this writing. Formats included blues, comedy and all Northwest rock bands (www.ibiquity.com), none of which would generally be considered commercially viable in a traditional analogue model.

In addition to programme diversity, digital transmission of course offers the enhancement of connectivity to the digital realm, of radio as gateway to web access and data services. Digital receivers can display song titles, news headlines and weather or traffic alerts, as well as interactivity with advertising. The commercial possibilities – such as the instant ability to push a button and buy a download of a song being played, or a product being advertised – makes digital transmission attractive to marketers. And, given the growing diffusion of third-generation mobile phones, software engineers have designed applications to allow users to listen to web streams of radio on their so-called 'smart phones' (Everhart 2009).

Do the potential programming, access and service benefits of digital radio, however, align with the 'touchstone value' of localism? In theory, the 'digital plenty' could allow for hyper-local, multicast channels that serve targeted geographic or ethnic communities. In practice, however, the early returns point to narrowcasting on secondary IBOC channels defined in terms of musical taste, such as the 'indie-rock' channels starting to proliferate on HD radio. Of course, satellite radio, as a national service, primarily construes 'localism' in terms of genre tastes, with the occasional exception of imagined communities such as the gay and lesbian audience served by Sirius' OutQ channel (www.sirius.com/outq).

Supervening social necessity?

While regulators and industry leaders invested years of energy and millions of dollars in competing digital radio systems, consumers have remained largely unimpressed. Presented with a variety of digital audio systems, the public has gravitated to iPods, smart phones and Internet audio, as indicated in the Bridge Ratings study. This reflects Winston's notion that there must be 'supervening social necessity' if a technological innovation is to be widely diffused (Winston 1998). Put simply, the innovation must serve some social need if it is to succeed; a commercial purpose alone is not necessarily sufficient. The rapid acceptance of early radio can be linked to satisfying the entertainment and information needs of a dispersed and growing population, as well as commercial needs for mass marketing (Lax 2003).

In contrast, where is the consumer demand for digital radio in the U.S.? What is the 'killer application'? Data from the Project for Excellence in Journalism indicates that the number of broadcasters switching to HD Radio peaked at in 2006 at 522 stations – the year HD became available to the public. The number of new HD stations fell 25% to 394 in 2007 (Radio 2008). While the data may reflect reluctance on the part of terrestrial station owners to shoulder the burden of conversion in the present difficult economic circumstances, the downward trend may also indicate that HD Radio has passed the peak in the bell curve of technology adoption described by Rogers (2003: 273). The PEJ study sees clouds on the horizon for satellite radio as well: public awareness of satellite radio has leveled off considerably since 2006 at about 60%. Interest in satellite radio remains flat as well, with only 3% responding that they are 'very likely' to subscribe in the next 12 months. Concurrently, 44% said cell phones are having 'a big impact on their lives', and today's smart phones are capable of receiving Internet streaming audio (including the streams of XM and Sirius), and analogue radio – but not HD Radio. In summary, it seems that analogue broadcast radio remains ubiquitous in the U.S., and hundreds of millions of receivers will work just fine for the vast majority of listeners for the foreseeable future.

What Levy wrote in the context of European digitization applies in the U.S. case: 'Technological change tended more often to lead to minor reforms of existing institutional structures…than to be used as an opportunity for a radical overhaul of either these institutions or hitherto shared policy objectives' (1999: 122). The lack of supervening social necessity, coupled with the absence of a government mandate to go digital (as in the case of television), crippled the dual system approach promoted by industry and regulators. Instead, consumers wearing ear buds are driving the 'radical overhaul'.

Notes

1. Personal communication from Skip Pizzi, 11 May 2009.
2. "The 'HD' in 'HD Radio' does not mean 'high-definition' or 'hybrid digital'. It is part of iBiquity's brand for its digital radio technology" (iBiquity 2009).

References

Ala-Fossi, M., S. Lax, B. O'Neill, P. Jauert and H. Shaw (2008), 'The future of radio is still digital – but which one? Expert perspectives and future scenarios for radio media in 2015', *Journal of Radio and Audio Media*, 15:1, May 2008, pp. 5–25.

Ala-Fossi, Marko and Alan G. Stavitsky (2003), 'Understanding IBOC: Digital technology for analog economics', *Journal of Radio Studies*, 10, pp. 63–79.

Andrews, E. L. (1992), 'Digital radio: Static is only between owners', *New York Times*, 6 May, p.D8.

Andrews, E. L. (1992), 'F.C.C. plan for radio by satellite', *New York Times*, 8 October, pp. D1, D17.

Andrews, E. L. (1995), 'F.C.C. backs digital satellite radio', *New York Times*, 13 January, p. D14.

Belsie, L. (1992), 'Digital audio broadcasting plays to global audience', *Christian Science Monitor*, 9 March, p. 9.

Berger, I. (2003), 'AM and FM play digital catch-up', *New York Times*, 25 December, p. G3.

Bouchard, G. (2007), 'HD radio technology trial in Canada', *CBC Technology Review*, 4, pp. 1–11, http://www.cbc.radio-canada.ca/technologyreview/pdf/issue4-trial.pdf. Accessed 26 January 2009.

Boyer, P. J. (1987), 'Under Fowler, F.C.C. treated TV as commerce', *New York Times*, 19 January, p. C15.

Bridge Ratings (2007), 'Competitive media usage overview and update', Bridge Ratings Press release, 23 May 2007, http://www.bridgeratings.com/press.05.23.07.CompMediaUse.htm. Accessed 26 January 2009.

Bunce, A. (1986), '"Digital audio" enhances radio sound and silence', *Christian Science Monitor*, 16 September, p. 21.

Desposito, J. (1999), 'IBOC proponents face final hurdle in quest for CD-quality broadcasts', *Electronic Design*, 47: 17, pp. 45–52.

Everhart, Karen (2009), 'Spare-time coder adds another way to hear NPR news on your cell phone', *Current*, 12 January, pp. 1, 8.

Feder, B. J. (2002), 'As digital radio stumbles, new products fill the gap', *New York Times*, 30 September, p. C2.

Feder, B. J. (2002), 'F.C.C. approves digital radio technology', *New York Times*, 11 October, p. C3.

Federal Communications Commission (1999), 'In the matter of digital audio broadcasting systems and their impact on the terrestrial radio broadcast service', MM Docket No. 99–325, FCC 99–327, http://www.fcc.gov/Reports/tcom1996.pdf .

Fleishman, G. (2005), 'Revolution on the radio', *New York Times*, 28 July, p. C11, http://www.jstor.org/action/emailSingleCitation?from=true&singleCitation=true. Accessed 9 February 2009.

HD Digital Radio Alliance (2005), 'Unprecedented radio-industry alliance will advance rollout of HD digital radio', Press release, 6 December 2005, http://www.hdradio.com/press_room.php?newscontent=16. Accessed 31 January 2009.

HD Digital Radio Alliance (2006), 'Radio companies kick off first phase of $200 million ad campaign for HD digital radio', Press release, 21 February 2006, http://www.hdradio.com/press_room.php?newscontent=23. Accessed 31 January 2009.

HD Digital Radio Alliance (2009), 'HD radio marketing toolkit', http://www.hdradioalliance.com/marketing_tool_kit.php. Accessed 31 January 2009.

HD Radio Alliance (2009), 'HD Radio – It's time to upgrade!', http://www.hdradio.com/find_an_hd_digital_radio_station.php. Accessed 27 January 2009.

Huntsberger, M. (2001), *Satellite Digital Audio Radio Service: Fundamental information on S-DARS*, Olympia, WA: Media and Communications Consulting.

iBiquity Digital (2003), '280+ stations in over 100 markets set to begin HD radio broadcasting', HD Radio Press release, 1 October 2003, http://www.ibiquity.com/press_room/news_releases/2003/83. Accessed 30 January 2009.

iBiquity Digital (2009), 'New buyer's guide', *HD Radio*, http://www.hdradio.com/buyers_guide.php. Accessed 30 January 2009.

iBiquity Digital (2009), 'Trademarks', http://www.ibiquity.com/about_us/trademarks. Accessed 30 January 2009.

International Center for Accessible Radio Technology (2008), 'Captioned radio broadcast to enable millions of deaf and hard-of-hearing to experience NPR's live coverage of presidential election for the first time', News release, 21 October 2008, http://I-cart.net.

Lawhorn, J. (2003), 'NPR initiates Tomorrow Radio Project', http://www.npr.org/about/press/030110.tomorrowradio.html. Accessed 27 January 2009.

Lax, Stephen (2003), 'The prospects for digital radio: Policy and technology for a new broadcasting system', *Information, communication and society*, 6, pp. 326–349.

Levy, David A. (1998), *Europe's digital revolution: Broadcasting regulation, the EU, and the nation state*, London: Routledge.

New York Times (1990), 'A dispute over radio technology', *New York Times*, 23 August 1990, p. D4.

New York Times (1991), 'Turf trouble in a digital day', *New York Times*, 2 June 1991, p. 115.

NPR Labs (2004), 'Tomorrow Radio Project announces stellar test results, declares victory in multi-channel HD radio research', Press release, 9 January 2004, http://www.npr.org/about/press/040109.tomorrowradio.html. Accessed 31 January 2009.

O'Neill, B. (2008), 'Digital radio policy in Canada: From analog replacement to multimedia convergence', *Journal of Radio and Audio Media*, 15:1, pp. 26–40.

Pizzi, S. (2008), 'Get ready for the new marketplace', *Radio World*, 32: 23, p. 42.

Pizzi, S. (2008), 'When evolution just isn't enough', *Radio World*, 32: 18, pp. 14–15.

Radio (2006), 'Canada rules to allow HD radio', *Radio: The Radio Technology Leader*, 27 December 2006, http://radiomagonline.com/digital_radio/eye_iboc/canada-allow-hd-radio/index.html. Accessed 31 January 2009.

Radio (2008), 'Mexico authorizes HD radio within 320 km of U.S. border', *Radio: The Radio Technology Leader*, 21 May 2008, http://radiomagonline.com/currents/mexico-hd-radio-us-border-0521/index.html. Accessed 26 January 2009.

Radio Act of 1927 (1927), *Columbia Law Review*, 27:6, pp. 726–733.

Radio World (2008), 'When do we stop calling it new media?', *Radio World*, 32: 19, p. 46.

Rogers, Everett M. (2005), *Diffusion of innovations; Fifth Edition*, New York: Free Press.

Shenon, Philip (2008), 'Justice Dept. approves XM merger with Sirius', *New York Times*, 25 March 2008, http://www.nytimes.com/2008/03/25/business/25radio.html.

Siepman, Charles (1946), *Radio's second chance*, Boston: Little, Brown and Company.

Stavitsky, Alan G. (1994), 'The changing conception of localism in U.S. public radio', *Journal of Broadcasting & Electronic Media*, 38, pp. 19–34.

Sterling, Christopher H. and John M. Kittross (2002), *Stay tuned: A history of American broadcasting*, Mahwah, NJ: Lawrence Erlbaum.

Telecommunications Act of 1996 (1996), 47 C.F.R. § 151 et seq.

Winston, Brian (1998), *Media technology and society*. New York: Routledge.

Chapter 7

Beyond Europe: Launching Digital Radio in Canada and Australia

Brian O'Neill

Eureka 147 was, as we have argued throughout this volume, a European technology designed within the very particular context of European public service broadcasting (see also Rudin 2006; O'Neill 2009). At the same time, the consortitum behind DAB technology had the ambition that Eureka 147 would become the world standard for digital radio. DAB was indeed the first such technological system to achieve standardisation at the European Telecommunications Standards Institute (ETSI), and be recommended as a global standard for digital terrestrial sound broadcasting by the International Telecommunications Union (ITU). Its early and highly successful demonstration at the convention of the National Association of Broadcasters (NAB) in the United States as early as 1991 was a further indicator of the strength of the system and the legitimacy of its claim to be a world leader (Hakanen 1991). However, the fact that it was a European-originated technology, ill-equipped to meet the different needs of the US radio industry, in addition to the potential disruptive effect it might have on a fiercely competitive market, meant that by 1992 DAB was off the agenda in the United States and never a serious contender for adoption (Ala-Fossi and Stavitsky 2003). Despite this, Eureka 147 has had success in many other countries across the globe and there are currently regular DAB services in Australia, China, Canada, Signapore and South Korea, with trial services in a host of other countries, including India, Indonesia, South Africa, Vietnam, Taiwan among others (WorldDMB 2009). This chapter examines two examples of DAB adoption beyond Europe: Canada, which had an early and close relationship with the development of the DAB standard, and Australia, which was the first country to formally adopt the DAB+ standard and in 2009 embarked on the largest such country-wide launch of the newer digital radio platform. Both examples offer comparable illustrations of national initiatives in supporting the roll-out of new digital radio technologies – one very recent, and one from the mid-1990s – and offer some important lessons for equivalent European experiences.

Eureka 147 in Canada[1]

Canada might be described as a pioneer of the first generation of digital radio broadcasting, and was among the first wave of countries who in the mid-1990s supported a policy of seeking a migration of all radio broadcasting onto a digital platform (Chouinard et al. 1994). In 1995 Canada formally adopted the Eureka 147 standard as the replacement technology for AM and FM analogue broadcasting (Canadian Radio-television and Telecommunications

Commission 1995; O'Neill 2007). This policy remained in place until 2006 when the regulator, the Canadian Radio-Television and Telecommunications Commission (CRTC), issued a revised digital radio policy, pursuing a more flexible, multi-platform approach to digital transition, reflecting the greater uncertainty surrounding future options for radio in the digital era (Canadian Radio-television and Telecommunications Commission 2006; O'Neill 2008).

More than just an enthusiastic early adopter of digital radio and Eureka 147, Canada was in fact closely involved in its development, and the government-funded Communications Research Centre played a central role in EBU technical committees involved in development of the Digital Audio Broadcasting system (Communications Research Centre 1997; Lavers 2003). As early as 1989, an ad-hoc advisory group convened by the private radio sector organization, the Canadian Association of Broadcasters (CAB), began to advocate the idea of a national strategy to implement the transition of the national broadcasting system from analogue to digital. The group with industry-wide representation – the public Canadian Broadcasting Corporation (CBC), private broadcasters represented by CAB, and representatives of the government Department of Communications – considered the various options available, and organized demonstrations of DAB in 1990 (Chouinard et al. 1994: 58). Following the success of these trials and the enthusiasm expressed across the radio sector for the project, a government Task Force was established to advise on the relevant technical, policy and regulatory aspects of introducing digital radio to Canada. It issued its report in 1994, outlining detailed plans for the relevant coverage and service issues, as well as making recommendations on the policy and regulatory implications (Task Force on Digital Radio 1995). An implementation body, Digital Radio Research Inc. (DRRI), (later Digital Radio Roll Out Inc.), was formed from the original consortium and became the official body mandated to promote digital radio in Canada. Industry Canada, the government department responsible for spectrum management, formally adopted Eureka as the standard for digital broadcasting in Canada, and allocated 40MHz of spectrum in the L-band range (1452–1492 MHz) for new broadcast services (Lavers 1993).

Canada's adoption of Eureka 147 was to some extent a radical and controversial choice given the different circumstances pertaining there (Chouinard et al. 1994: 60). In contrast to the European situation where public broadcasters were the key proponents of the new system, the Canadian Broadcasting Corporation was a part of but by no means the leading advocate for digitalizing radio (O'Neill 2006). Broad support from the private radio sector and co-ordination by its representative organization was a key feature of the Canadian approach. The decision to develop an entirely new band for digital radio using L-Band spectrum was also in contrast to the Band III VHF spectrum preferred in Europe. Canada's VHF spare capacity was very limited however. L-band, on the other hand, offered a new platform from which to launch digital radio and which, in the Canadian view, would be the least disruptive for the existing radio market (Task Force on Digital Radio 1995: 18). Accordingly, Canada lobbied strongly for L-Band development, and in 1992 won WARC approval for L-Band as the only radio spectrum allocated exclusively for digital radio transmission on a worldwide basis. Internationally, actual implementation of digital radio on L-Band continued to be the less preferred option given its lower propagation characteristics and added costs.

An integral feature of the Canadian plan was its proposed integrated mix of terrestrial and satellite delivery for which L-Band was deemed the best approach (Communications Research Centre 1997). This, however, placed it in conflict with the United States where the military had reserved it for systems testing, blocking its use for other purposes. Subsequently, the United States reached an agreement with Canada that the latter would restrict L-Band DAB to terrestrial broadcast in order to avoid interference (Barboutis 1997). But the Canadian satellite plan was now compromised and placed DAB transmission on a more expensive and challenging terrestrial-only footing. The United States' opposition to the use of L-Band for radio broadcasting stemmed from the threat that it posed to the established market for AM and FM radio (Ala-Fossi and Stavitsky 2003). Unlike in Canada where it was proposed to move all existing AM and FM stations onto the L-Band, the crowded and highly competitive US market ruled out any such migration. Developing a new band for digital radio would create unacceptable competition. Accordingly, United States' radio interests developed the alternative 'in-band, on-channel' approach as a technical solution to enable digital radio transmission within existing bandwidth and without recourse to L-Band or satellite transmission (Rathbun 2000). By contrast, the Canadian approach was designed to complement existing radio interests: each AM and FM broadcaster would have an equal opportunity to compete in the new digital environment. It was hoped that the phased transition to L-Band outlined in the regulator's 1995 digital radio policy (Canadian Radio-television and Telecommunications Commission 1995) would create an optimal digital radio system, sufficient to allow the shut down of analogue services in due course.

A further feature of the Canadian policy was the requirement that digital radio offer demonstrably superior quality, including audio quality of 256kbs (thereby limiting each multiplex to a maximum of five programme services), as well as ensuring there would be no interference with existing AM/FM services. In the agreed licensing framework, all existing AM and FM licensees were allocated frequencies in the 1452–1492 MHz L-Band, and an allotment plan was developed in each of the major metropolitan areas to allow the digital service to match as closely as possible the coverage of existing stations. This involved defining DAB coverage around the largest FM station within any given market and grouping up to five existing FM stations into a single multiplex. Replacement of wide area AM stations was restricted to the largest equivalent, though smaller, FM coverage area. As a replacement technology, stations were licensed only to simulcast existing services and were not permitted to offer new or additional services. Equally, no new licences were to be offered for the duration of the transition nor would any new operators be enabled to enter the market until a full migration of existing services had occurred.

The pioneers of digital radio in Canada could claim considerable optimism for its prospects during the initial inception phase. The groundwork and development for DAB in Canada was described as a textbook case of co-operation among the many interests involved (Chouinard et al. 1994: 79). The relatively small group at DRRI who pioneered and championed the cause of DAB ensured that the technology had been perfected and standardized, the necessary spectrum had been obtained and a solid foundation was in place for large scale implementation (Edwards 2001). With sufficient marketing, public information and availability of receivers at a reasonable cost, it was widely believed that consumers would be clamoring for DAB (Bray 2000).

An official launch of digital radio in Canada took place in 1999 at the convention of the Canadian Association of Broadcasters, and a steady roll out of stations with DAB services proceeded in key metropolitan areas. Within a short period, there were 57 stations broadcasting in DAB, reaching 35% of the population, some 10 million listeners in Toronto, Montreal, Windsor and Vancouver, with a further launch of DAB services in Ottawa to follow (Bray 2000). By 2002, the CAB's vice president of radio could confidently claim Canada's emergence as a world leader in digital radio (Cavanagh 2002: 30). A major boost to the marketing of DAB was the announcement by General Motors of Canada of its plans to install DAB receivers as standard equipment in its vehicles for the 2003 model year. Also in that year, DRRI commissioned an engineering study to extend the national coverage of DAB in a series of corridors between the major metropolitan centres. Significant progress also appeared to be underway in receiver availability: Radio Shack Canada announced it would carry a range of home and portable DAB consumer products across its stores. The development of a new DAB chip by Texas Instruments also promised greatly reduced prices for receivers, and the first DAB/FM personal portable below the psychologically all-important $100 became available.

What went wrong for DAB in Canada?

Despite these positive early indications, DAB in Canada, as elsewhere, did not live up to expectations or develop as the mass consumer technology as expected (Lavers 2003). By 2006, there were officially 73 licensed DAB stations in Canada or 62 that were fully operational, primarily in the main cities of Toronto, Vancouver, Montreal and Ottawa. Coverage was estimated to reach eleven million potential listeners. However, listenership for all stations was low and was not even monitored by official audience measurement. DAB receivers were never readily available either for home or car use largely due to the lack of L-Band receivers. Industry professionals soon became disillusioned and regarded the investment in time and resources since 1995 as unproductive and wasteful. By the time digital radio policy came to be reviewed in 2006, the objective of migrating all radio to DAB was regarded as misguided. The reasons for the failed deployment are familiar from other markets around the world and call into question many of the founding assumptions guiding the first phase of digital radio.

In the first place, it is clear that there was a very poor consumer response to the development of DAB in Canada, and at no time within the period 1995 to 2005 could it be said that digital radio firmly took hold. There was poor awareness of the service and, indeed, even of the existence of the new technology or its potential benefits for radio listening. There were particular difficulties with the supply of receiver equipment and it was erroneously assumed that a range of equipment would follow with the adoption of DAB in Europe. Despite DRRI's efforts to create awareness of DAB and its benefits, the fact that receivers were largely unavailable or difficult to source proved extremely damaging to the prospects of an early take up of DAB. Initial costs of around $2000 for high-end consumer receivers gave DAB an elite image that proved difficult subsequently to shake off. Lower cost receivers, once they were available, performed poorly, adding further difficulties to any potential increase in supply of receiver equipment.

With poor availability of receivers, and in many instances poor quality of what was available, the much-heralded enhanced features of the digital radio listening experience proved to be unattainable or far below expectations. The assumption that the promise of enhanced CD-like audio quality would be the unique selling feature of the new technology proved unfounded in Canada as elsewhere. It was also the case that many of the promised additional services did not arrive either, with most stations simply offering simulcasts of their analogue services. Despite initial enthusiasm for the possibilities of data services, with few exceptions none really materialized.

The failure to gain the support of the automotive sector was also a blow to DAB in Canada. DRRI's success in getting a commitment from General Motors Canada for installation of DAB receivers in its 2003 models proved short lived when difficulties emerged over supply of equipment, and segmenting production for those areas of the Canadian market where DAB was available (Lavers 2003). In the absence of a timetable for the roll-out on a national level of DAB, General Motors withdrew their support for a standard fit out of DAB and like the rest of the sector began to adopt 'a wait and see' approach. General Motor's subsequent shareholding interest in XM satellite radio in the United States, in due course a competitor against DAB and conventional radio in the Canadian market, further impeded any long-term interest in the platform.

Despite the initial enthusiasm and strong support by private broadcasters for the DAB project in the early 1990s, its failure to take off either internationally or in the Canadian market led to a gradual cooling of enthusiasm if not outright withdrawal of support. Doubts began to emerge over the decision to adopt Eureka 147 once it became clear that a different approach was to be pursued by the United States. Canada's decision to adopt DAB was made in the knowledge that this would not be followed in the United States, and the subsequent development of IBOC or HD-Radio placed Canada and the United States at odds, with radically different approaches to digital radio broadcasting (Federal Communications Commission 2007). It was assumed that as radio was primarily a local medium, use of incompatible systems either side of the border would not prove too significant. Experience has shown, however, that it is difficult for Canada to pursue a different course to its near neighbour and, unquestionably, the adoption of IBOC in the United States contributed to the growing unease among industry members in Canada about the wisdom of their DAB policy (see Stacey 2007). In spite of the fact that Eureka 147 DAB was acknowledged to be the technically superior system (Bouchard 2007), many industry executives came to accept that successful implementation of IBOC in the United States would eventually lead to its adoption in Canada. Further objections to DAB began to be raised following estimates of the investment required to extend its coverage, and the proposal to establish transmission corridors between major metropolitan centres proved to be a prohibitively expensive proposition.

This cooling of support from the industry for DAB was reflected in the relatively low profile adopted by CBC, the national public broadcaster (O'Neill 2006). In contrast to the European situation, where public broadcasters played a leading role in the development and the roll out of the technology, CBC has been a participant rather than a leader in DAB (Patrick et al. 1996; Galipeau 2003). A member of the original Task Force for the Introduction of Digital Radio and a 50% partner in DRRI, CBC was an active and equal participant in industry efforts to steer the sector towards the digital domain. However, CBC was not a champion of DAB and did not develop any dedicated digital-only services or lend DAB any particular promotional support. DAB was effectively co-opted as one of a number of options in an overall new media strategy which included the Internet, subscription digital audio services via cable and, more recently, satellite broadcasting – for which it did develop new services (O'Neill, 2006). In truth, the downsizing of CBC's engineering division (Lavers 2006) reduced its capacity for anything other than programme-driven priorities, and while CBC now experiments with newer applications such Digital Multimedia Broadcasting or DMB using DAB technology, its interest in digital terrestrial radio *per se* has waned considerably.

One further effort to launch DAB in Canada was attempted in 2006 when the Canadian company CHUM Ltd. proposed a subscription radio service across Canada on a digital terrestrial DAB network to compete with the imminent launch of satellite radio. Satellite radio had made a high profile entry into the Canadian market in 2005 when both XM and Sirius platforms were licensed to operate their subscription service under revised Canadian broadcasting regulations (O'Neill 2007; O'Neill and Murphy in press).

CHUM proposed the establishment of a national terrestrial DAB network providing 50 channels initially, subsequently growing to 100 channels, for a monthly subscription fee of $9.95. Controversially, whereas the bulk of the satellite's music service of over 100 channels was not subject to the normal Canadian content regulations, CHUM's proposal as a terrestrial service was licensed under the normal content rules for all Canadian broadcast services. CHUM signed a technology agreement with equipment manufacture RadioScape for specially designed receivers that could also be used to pick up regular, non-subscription DAB channels, thereby re-launching DAB in Canada. The proposal included using existing DAB allocations for AM and FM replacement, as well as reduced bit-rates of 128 kilobits per second to achieve the density required for a 50 channel service. Adoption of DAB+ in the future was also suggested, with consideration to be given to 'more advanced codecs providing an approximate doubling of spectral efficiency' (Pizzi 2004).

CHUM's proposal never got beyond the planning stage, and while its application was approved, the fact that its satellite rivals could offer greater choice without Canadian content restrictions completely undermined its business case. While the intention may have been to restart debate about Canada's digital radio policy in offering a pro-Canadian solution to new competition, the response of the regulator was to effectively let the market decide. The fact that no special protection was afforded to CHUM's 'Canadian' proposal effectively spelt the end of the strategy of migrating the industry onto the Eureka 147 platform.

Summing up what had been an unproductive ten-year period since the initial introduction of DAB into Canada, one radio executive candidly remarked:

> I believe it was a waste of time and money and we are still sitting here with nothing. I never understood (it). I said from day one there's no indication that consumers want replacement technology. They don't see our signal being as bad as we think they think it is. And I don't think we ever researched it correctly. In terms of our plan which was always to put our existing stations on a new platform and transition – waste of time, money and no demand. (G. Slaight, personal communication, August 12, 2005)

Towards a new digital radio policy

The Canadian regulator, the CRTC, recognizing the stalled state of digital migration, issued a revised digital radio policy in December 2006 (Canadian Radio-television and Telecommunications Commission 2006). The notice acknowledged those factors which had acted as blockages to the roll out of digital transmission such as poor availability of receivers, the use of L-Band, the development of IBOC and the growing interest in satellite radio and other forms of digital distribution. The lack of additional digital content was also agreed to be a major constraining factor in that consumers were essentially being encouraged to buy new receivers to receive the same content. Of particular relevance to Canada was the fact that digital radio was only available in the major markets. The proposal to roll out digital transmission in the main traffic corridors between major cities had proved prohibitively expensive, and as a consequence the automobile industry had switched its support to satellite subscription radio (CRTC, 2006: 6).

The revised policy, developed in consultation with the sector, was intended to provide a new regulatory environment for digital radio broadcasting and new options for the radio industry. The industry had confirmed in its submissions to the Commission that the original 1995 policy had failed. The CBC argued that the transitional policy should be abandoned, and that 'the future of DRB in Canada will not be that of a replacement technology, but as a technology that will co-exist with the existing analogue radio services' (CBC/Radio-Canada 2006). The CAB, representing private radio, argued that it was 'simply not realistic to assume that a successful digital transition will be no more than the replacement of the existing business with minor additions and adjustments. Nor does digital transition necessarily mean the destruction of the old business and the creation of a new one' (Canadian Association of Broadcasters 2006). A transition to digital in the Canadian context would only be successful on the basis of commercially viable propositions that included new content, affordable receivers, promotion and competitive technical features. Eureka 147 would continue to be part of the equation but so would its variants, such as multimedia broadcasting, options for IBOC, Internet distribution and technologies for distribution to hand-held mobile devices.

Therefore, the most important provision of the revised digital radio policy was for a digital 'new service model', overturning the replacement strategy based on simulcasting AM and FM services on L-Band (CRTC 2006: 38). Licensees under the revised policy were now free to develop whatever broadcast services they saw fit, subject to the same regulatory provision as existing FM services, though with greater flexibility promised for specialty services. Radio station owners continue to have privileged access – each guaranteed a digital licence for every analogue licence held –with the aim of facilitating licence holders to build additional services in L-Band while maintaining their FM and AM holdings. The new service model explicitly recognized the co-existence of analogue and digital broadcasting into the future, acknowledging that FM would continue to show robust growth, with only the long-term future of AM being in any doubt.

From a technical point of view, the limit on five stations sharing a 1.5 MHz L-Band channel or multiplex remains in place though the policy suggested that DAB+ could be adopted in the future (it was at the time undergoing standardization). In addition, however, IBOC technologies, subject to satisfactory testing for any potential interference to other stations, would be permitted for licensing. Technologies which re-used existing analogue spectrum such as IBOC or DRM had advantages and would also be considered for licensing. Finally, the revised policy supported testing multimedia broadcasting options (such as DVB-H and DMB technologies) to deliver a mix of audio, video and related data services in the L-Band subject to sufficient spectrum being made available.

In advance of the policy, Canadian broadcasters had called for a flexible, multi-option plan that would enable them to compete with the new digital audio services now widely available to consumers. They sought a flexible approach to digital radio conversion that would support supplementary services rather than replace analogue broadcasting (Stacey 2007). The revised digital policy provided broadcasters with everything they required and gave licensees every freedom 'to propose the technology or technologies they believe will be the most appealing to the listening public' (Arpin 2007). Somewhat ominously, the Commission added that no guarantee of success could be offered given that many of the factors that had led to the current stalled-state of Canadian digital radio, including poor availability of receivers, remained in place.

The industry response to the new policy was communicated at a round table convened by the CRTC in 2007. At this event, industry representatives gave a very cautious assessment of IBOC's potential for Canada, having found a number of technical deficiencies compared to analogue FM coverage (Bouchard 2006; Lehane 2007). Likewise, there was little enthusiasm for a repeat of the simulcasting on digital that had previously been so unsuccessful. Instead, the consensus view was that multimedia broadcasting offered the greatest potential for a 'new service model' for Canada and the single best opportunity for digital radio implementation.

Demonstrations of DMB integrating radio, mobile video and data-casting, and converging on devices such as cellular telephones were proposed by the industry as the optimal digital radio solution. The proposal included an ambitious plan for 70 Single Frequency Network (SFN) sites across Canada, aiming to reach approximately 60% of the population in three years at a total capital cost of CDN $106 million, using DMB in the L-Band to support multimedia broadcasting to handheld devices. A consortium approach was proposed as the best framework to build out the necessary infrastructure, whilst content partnerships between companies, as well as with telecommunications, were envisaged as the way forward.

Given that the proposal involved a substantial change of use for L-Band spectrum initially granted for digital radio transition, the spectrum regulator Industry Canada intervened (Industry Canada, 2007), and suspended any further development of services pending the outcome of a consultation on the future use of L-Band. Suggesting that broadcasters might not be the only parties interested in the spectrum and that alternative means of allocating it including auction might be considered, Industry Canada added that the consultation would determine the most appropriate allocation for the band and the means of making the spectrum available for potential users. At the time of writing, no further progress has been made on the future use of L-Band and it would appear that once more Canada's digital radio policy has stalled.

Digital radio in Australia

There are many points of comparison between Australia and Canada in terms of digital radio implementation. There are obvious geographical similarities in that both countries occupy a large contintental land mass requiring extensive and powerful transmission systems to ensure universal coverage. The centres of population in both are concentrated in the major metropolitan areas. Radio markets in Canada and Australia are also comparable: as in Canada, Australia's economically vibrant media system has a strong tradition of public broadcasting, but privately-owned broadcasting has the dominant share of listening and viewing. Ownership of the media in both countries is also highly concentrated and, following changes to ownership rules in both, there has been considerable consolidation of the industry over the past five years. Currently 80% of radio stations in Australia are in the hands of twelve radio networks (Nielsen Media Research 2009). There are currently about 261 commercial radio stations on air in Australia and commercial radio accounts for 70% of listening. There are approximately 37 million radio sets across the country, every household has at least one radio and on average have 5.1 sets, 99% of all cars having a radio and almost 8 in 10 Australian's listen to commerical radio every week (Commercial Radio Australia 2009). The public broadcaster, the Australian Broadcasting Corporation (ABC), runs national and local public radio and TV stations as well as Australia Network, a TV service for the Asia-Pacific region. The other main public broadcaster is the Special Broadcasting

Service (SBS), whose radio and TV networks broadcast in many languages. There is also a strong community radio sector with 230 stations across the country and a further 1500 low power FM and temporary community stations.

Digital radio was first considered in Australia in the late 1990s. A policy for the introduction of digital television in 1998 also included a plan for digital radio which never proceeded to implementation (Given 2003: 149). A report to the government in 1997 had recommended Eureka 147 operating in the L-Band, taking Canada's lead, while also exploring the possibility of VHF spectrum in Australia's crowded metropolitan areas, as well as the potential of satellite technology for national coverage. The implementation model proposed also followed the Canadian approach in recommending automatic access for all national and commercial broadcasters in a transitional or developmental phase during which they would simulcast on analogue and digital platforms in order to develop the new technology and experiment with new services. Incumbents would get priority access and a moratorium on any new commerical competition for a period of years. New entrants would in due course get access to frequencies to further develop the market for digital radio services, and at some point in the future, analogue switch-off would follow.

While interest in digital radio was strong and considerable planning was put into how best to develop it in Australia, issues over availability of spectrum and capacity for all the radio interests involved served to delay any move to implementation. Channel 9A – the TV-band VHF frequency between 202MHz and 209MHz – was granted by the regulator first for trials of digital radio broadcasting and subsequently for licensing digital radio services. While in most metropolitan areas there was sufficient capacity within this band for three multiplexes, a hybrid solution using both Band III and L-Band would be required to enable sufficient capacity for all interested licence holders. This added greatly to the cost of transmission and would place additional burdens on radio stations converting to digital, and as a result digital radio fell off the agenda for a period as broadcasters weighed up the costs invovled (Given 2008). At the same time, new audio services were being developed without the need for a digital broadcasting network, such as ABC's new DiG or digital radio service available through digital TV and via the web. Further expansion of commerical, community and low power FM services raised the question as to whether additional digital choice was needed.

The fact that DAB in Australia did not launch as originally planned did have the benefit of enabling the industry to re-evaluate its approach to digital radio. With the rapid adoption of digital television, broadband Internet and digital audio devices, the resolve of the sector to develop its presence within the digital environment was as strong as ever, and Commercial Radio Australia (CRA), the representative organisation of private radio in Australia, continued to lobby strongly for its introduction. Legislation in 2006 prepared the ground for a new launch to digital radio, this time earmarked for 2009, taking into account varying international experiences such as the United Kingdom and Canada, new technology developments, in particular better compression technologies, and a consolidated industry view of what was required to make digital radio a success.

The government's policy framework included a number of important developments (Coonan 2005a). For one, digital radio was defined as a *supplementary* rather than a *replacement* technology. Based on the ten years of experience of digital radio deployments elsewhere, it was acknowledged as highly unlikely that all analogue radio services could be converted to digital or indeed that digital radio might ever be a complete replacement for analogue broadcasting. It was observed that no country that had introduced any of the digital radio platforms had done so with a realistic expectation of analogue shutdown (Coonan 2005b). Given that Australia did not have the problem of legacy technologies as for example in the United Kingdom, with a large installed base of DAB receivers, a new digital radio launch in Australia could now benefit from the very latest new technology developments – specifically DAB+ using the improved and more efficient AAC coding – thereby increasing capacity and quality of services.

The framework for digital radio implementation now proposed included a moratorium of six years during which incumbent commercial broadcasters would have priority access to digital spectrum, subject to conditions attached to roll-out and coverage requirements for digital transmission. In contrast to the UK where one company, Digital One, controlled the commercial multiplex, Australia's framework gives commercial stations and larger community stations the right to jointly own and manage the multiplex infrastructure and hold the associated spectrum licences. In addition, access rules are established to ensure that multiplexes are operated in a transparent and non-discriminatory way. Finally broadcasters would not be constrained by being required to simulcast their existing services, and would be free to develop whatever new services or enhancements they felt appropriate to encourage the take up of digital radio.

Australia's 2009 digital radio launch

The launch of digital radio in Australia in 2009 represents a critical test of new approaches to digital radio broadcasting. It is the first formal launch of the new DAB+ platform, with its added features and superior quality, and its success may determine future deployments in countries such as China and India. Drawing on the many experiences of previous efforts to launch digital radio in the marketplace, Australia has had the advantage of learning the lessons from numerous previous examples of what is required to incentivize the industry and to attract consumer interest. To date, the approach adopted by Commercial Radio Australia, co-ordinating on behalf of the sector, has been held up as a textbook case of co-operation between industry interests – private and public – and the government, and offers the best opportunity yet for a systematic roll-out of a new, digital terrestrial radio broadcasting platform. The fact that it represents the first test case for the DAB+ platform is also highly significant for the future of the Eureka family of technologies.

According to CRA, one of the key lessons to be learned from the UK experience is the importance of offering not just new content but also a new experience for radio listening (in Given 2009). For this, the added capacity and multimedia capabilities of DAB+ are central.

The UK experience of providing new content and added choice, Australian broadcasters point out, has been a partial success, but yet in ten years has only achieved 10% of overall radio listening. A further lesson is that in launching the new digital radio service, digital radio needs to offer as comprehensive a range of radio listening as possible, offering a 'one stop shop' for all current AM, FM and new digital services. Sensing that the attachment to a familiar and much-loved medium must be preserved at all costs, a carefully crafted marketing plan based around an enhanced radio listening experience has been undertaken. The tag line for the CRA's $AUS10 million digital radio promotional campaign – 'its radio you know and love, plus more' – seeks to convince listeners that the new technology will host all current favourite services, with added features and enhancements. Among the new features being promoted are those of superior quality, 'crystal clear' sounds that will provide an immediate sonic upgrade for the 48% of Australian radio stations operating in the AM band; electronic programme guides and other visualization add-ons such as scrolling text, animated logos and product photos; as well as the pause and rewind features that are available in more sophisticated sets.

The switch-on date for the new digital services was initially planned for May 1st 2009, with the aim of having the first five markets in the mainland capital cities of Sydney, Melbourne, Adelaide, Brisbane and Perth launch simultaneously. The government has required that digital radio services must be on air in the state capital cities (excluding Hobart) by July 1 2009, and all existing national, commercial and wide-area community broadcasters in each city be able to broadcast digitally. It is unlikely that digital radio will be extended to regional areas until more broadcast spectrum becomes available following the turn-off of the analogue television network in 2013 (ABC 2009). The formal launch was later delayed to August 2009, in part due to delays and funding difficulties at the public broadcaster ABC in installation of its transmission equipment. Instead, May 1st was used as a date for test signals being switched on, to be followed subsequently by a saturation advertising campaign. A key priority for the project, regarded as key to its success, has been to maintain its unified approach and reinforce the fact that all broadcasters are launching on the new platform with existing and new services. The other essential ingredient has been to ensure sufficient supply of receivers for what is a new digital format, likewise seeking to avoid mistakes of earlier initiatives which were severely undermined by poor retail support and consumer awareness.

Despite the fact that the Australian launch has been one of the most extensive and best planned launch campaigns to date for digital radio, a number of issues persist which pose further challenges to its long-term sustainability. A problem familiar from the example of Canada is the fact that the launch is restricted to the major metropolitan areas, initially the five capital cities of Sydney, Melbourne, Brisbane, Adelaide and Perth. Once outside of the coverage area, listeners need to retune to AM and FM. While an option for satellite transmission has been considered and the spectrum reserved for its future development, the costs are likely to be as prohibitive as elsewhere (Broadcast Australia 2005). Likewise, it has been conceded that DAB is not suited to full national coverage and that an alternative digital platform such as Digital Radio Mondiale will be needed in the future (Coonan 2005b). Therefore, for the moment, digital radio will be restricted to major radio markets in cities, but crucially will lack the universal coverage that existing analogue services have. Within cities, concerns have been expressed about the quality of the spectrum that has been allocated for digital radio use, and that problems in field signal strength have already posed difficulties in ensuring quality of reception in buildings (Braue 2008). This is not an ideal situation, as the CRA readily admit, when trying to convince the public of the superiority and quality of the service on digital.

Furthermore, while spectrum planning and the organization of Australian multiplexes has ensured that all public and commercial broadcasters are represented on the new platform, there is insufficient capacity to mirror the full range of services that currently exist on analogue. In particular, the large community radio sector has struggled to gain sufficient spectrum, even for some of its city-wide services, and has expressed its unhappiness with the provision for community radio (Given 2009: 4). All other stations have been allocated a minimum of 128kbs, enough for at least two FM quality services plus data services. Stations are free to use their bandwidth as they see fit and are not, as in many previous launch schemes, required to simulcast their existing services. Given the higher audio quality of DAB+ at lower bit rates, talk radio, for instance, will have extensive capacity to offer new channels, whereas for music, the trade-off between quality and additional content will continue to be a factor.

A further question for Australian digital radio is whether the broadcast platform can be sufficiently appealing to consumers against a host of other competing digital audio services, including MP3 players, mobile communication services and the Internet. Is its digital radio offering, in other words, too late to be able to successfully compete in the digital audio market place where rival services have had a lead time of ten years to develop and mature? Commentators point out that many of the features that digital radio promises, even in its new DAB+ version, are currently available in existing products, analogue and digital. The wide availability of high quality FM reception, RDS tuning and a host of personal media players mean that there may be little incentive for consumers to invest in a service that may be of only marginal benefit. Industry interests concede that digital radio may be a long-term project, relying on a strategy of embedding services with younger audiences on the understanding that a sustainable level of adoption may take many more years (Given 2009). Despite this, the Australian radio industry has pressed ahead with strong support from Australia's leading commercial radio operators and the public broadcaster, confident that the new technology of DAB+ and the potential for success of its new brands and services provides the right way forward for the development of the medium.

Conclusion

Lessons learned from across Europe and beyond have contributed to a growing understanding of the ingredients required for a successful launch of digital radio in the twenty-first century. To be effective, industry experts argue that digital radio needs to offer: a) a strong consumer proposition to include attractive new content, probably involving multimedia, a good range of appealing receivers and guaranteed good reception; b) it requires an integrated approach from broadcasters – who understand the benefits of lower transmission costs and new channel capacity; and finally, c) it needs an encouraging framework from governments and regulators (Howard 2008). Many of these elements were absent from the first generation of digital radio roll out, of which the Canadian experience was a classic example, and consequently digital radio there has a low priority among broadcasters and has lost momentum within Canadian media development strategy. By contrast, Australia has reaped the advantage of learning from others' experiences and, despite having hesitated, is in a better position to move forward with a newer technology and a more realistic assessment of the role of digitalization in the development of the medium.

Australia's role as a pioneer of the new generation of terrestrial digital radio broadcasting is important for other countries around the world considering new deployments of digital radio, and an important test case for Europe in particular. With the major radio markets of Germany and Italy proposing new launches of DAB+ from 2010 on, in addition to the launch of digital services on T-DMB in France, Australia's much admired example of industry coordination and strategic thinking will be followed closely. It is clear from the examples of both Canada and Australia that replacement of the analogue transmission network is no longer the primary motivating factor of digital radio deployment, and that a full migration, if it is to occur at all, is a very long-term objective. It is also recognized that the uptake of digital radio is an inherently slow process and that it will be many years before the success of current strategy or investment is known.

Both Canada and Australia also illustrate new thinking about the 'additionality' involved in a digital radio strategy. Canada's new service model or the supplementary nature of many of the new Australian services illustrate again the age-old maxim of content being king, and suggest that for digital radio to succeed it will need exciting new content not available on other media. Whether such content can be produced economically, without a proven business model for digital radio as a distribution platform, remains to be tested. Canada's approach suggests that digital radio will be a niche service, possibly paid for by subscription, whereas in Australia the emphasis has been on free-to-air radio, augmenting existing services with new content and new functionality. The underlying rationale articulated by Australia's radio industry, reiterating earlier arguments for the development of the technology, was that it was inconceivable that radio would be the only mainstream broadcasting platform to remain analogue-only (Coonan 2005a). Whether the investment in DAB+ proves to be the correct choice will be the subject of much debate in years to come.

Note

1. This section draws on research previously published in the *Canadian Journal of Communication* (O'Neill 2007) and the *Journal of Radio and Audio Media* (O'Neill 2008). Original research for this chapter was supported by a grant from the Ireland Canada University Foundation.

References

ABC (2009), 'What is digital radio?', http://www.abc.net.au/radio/digital/. Accessed 1 June 2009.

Ala-Fossi, M. and A. G. Stavitsky (2003), 'Understanding IBOC: Digital Technology for Analog Economics', *Journal of Radio Studies*, 10:1, pp. 63–80.

Arpin, M. (2007), 'Speech to the Annual General Meeting & Conference of the North American Broadcasters Association', *Annual General Meeting & Conference of the North American Broadcasters Association*, Mexico City, Mexico, 6 March 2007.

Barboutis, C. (1997), 'Digital audio broadcasting: the tangled webs of technological', *Media Culture & Society*, 19: 4, pp. 687–90.

Bouchard, G. (2006), 'First Canadian Encounter with the New Radio Transmission Technology', *CBC Technology* Review, 2, pp. 1–12.

Bouchard, G. (2007), 'HD Radio™ Technology Trial in Canada', *CBC Technology Review*, 4, pp. 1–11.

Braue, D. (2008), 'Internet killed the (digital) radio star', *cnet Australia*, http://www.cnet.com.au/internet-killed-the-digital-radio-star-339289967.htm. Accessed 1 June 2009.

Bray, D. (2000), 'A little DAB will do ya', *Broadcaster,* June 2000.

Bray, D. (2000), 'Digital Radio: A Numbers Game', *Broadcaster,* September 2000.

Broadcast Australia (2005), 'Submission By Broadcast Australia In Response To The "Introduction Of Digital Radio Issues Paper"', December 2004, www.archive.dbcde.gov.au/__data/assets/pdf_file/0020/26075/Broadcast_Australia_public_submission.pdf. Accessed 1 June 2009.

Canadian Association of Broadcasters (2006), 'A Submission to the Canadian Radio-Television and Telecommunications Commission with respect to Broadcasting Public Notice CRTC 2006-72'.

Canadian Radio-television and Telecommunications Commission (1995), 'Public Notice CRTC 1995–184: A Policy To Govern The Introduction Of Digital Radio', Ottawa, Canada.

Canadian Radio-television and Telecommunications Commission (2006), 'Broadcasting Public Notice CRTC 2006–160. Digital radio policy', Ottawa, Canada.

Cavanagh, R. (2002), 'The future is sound with DAB', *Broadcaster,* January 2002, p. 30.

CBC/Radio-Canada (2006), 'CBC Submission, Review of the Commercial Radio Policy – Broadcasting Notice of Public Hearing CRTC 2006–1', http://www.cbc.radio-canada.ca/submissions/crtc/2006/BPN_CRTC_2006-1_CBCSRC_150306_e.pdf. Accessed 1 June 2009.

Chouinard, G., F. Conway, W.A. Stacey and J.R. Trenholm (1994), 'Digital radio broadcasting in Canada – A strategic approach to DRB implementation', *EBU Technical Review,* Winter.

Commercial Radio Australia (2009), 'Radio Facts', http://www.commercialradio.com.au/index.cfm?page_id=1007. Accessed 1 June 2009.

Communications Research Centre (1997), 'Digital Radio Broadcasting and CRC', http://www.crc.ca. Accessed 1 June 2009.

Coonan, H. (2005a), 'Framework for the introduction of digital radio', Media release, http://www.minister.dcita.gov.au/coonan/media/media_releases/framework_for_the_introduction_of_digital_radio. Accessed 1 June 2009.

Coonan, H. (2005b), 'Speech to Commercial Radio Australia Conference', http://www.minister.dcita.gov.au/media/speeches/digital_radio_-_commercial_radio_australia_conference. Accessed 28 May 2009.

Edwards, S. (2001), 'Spoiled For Choice Issues and Options For Digital Radio', Speech to *Radio, Television and the New Media*, Canberra, Australia, Australian Broadcasting Authority, 3–4 May 2001.

Federal Communications Commission (2007), 'Digital Audio Broadcasting Systems And Their Impact on the Terrestrial Radio Broadcast Service', http://www.fcc.gov/mb/policy/dab.html. Accessed 1 June 2009.

Galipeau, C. (2003), 'New Media at CBC', Presentation at *Publishing Across New Media Platforms*, Simon Fraser University, Vancouver.

Given, J. (2003), *Turning off the television: broadcasting's uncertain future*, Sydney, Australia: UNSW Press.

Given, J. (2008), 'Broadcasting. Has radio blown the future?', *Creative Economy Online*, http://www.creative.org.au/. Accessed 1 June 2009.

Given, J. (2009), 'Content to the Different', *The Herald Sun*, 5 March, p. 4.

Hakanen, E. A. (1991), 'Digital audio broadcasting – promises and policy issues in the USA', *Telecommunications Policy*, 15:6, pp. 491–496.

Howard, Q. (2008), 'The Attributes of Digital Radio', Presentation at *EBU Digital Radio Conference 08*, Cagliari, Sardinia, 18–19 September.

Lavers, D. (1993), 'Canadians shift L-band DAB into gear', *Broadcasting*, 123:4, pp. 116–18.

Lavers, D. (2003), 'Communications Research Centre', *Broadcast Dialogue*, November 2003.

Lavers, D. (2003), 'What now for DAB in Canada?' *Broadcast Dialogue*, June 2003.

Lavers, D. (2006), 'CBC Technology – A Phoenix Rising', *Broadcast Dialogue*, March 2006.

Lehane, S. (2007), 'Digital radio: state of the industry – crtc opens doors to multiple digital radio standards', *Broadcast Dialogue*, April 2007.

Nielsen Media Research (2009), 'All homes in Australia have at least one radio', http://www.nielsenmedia.com.au/industry.asp?industryID=13.

O'Neill, B. (2006), 'CBC.ca: Broadcast Sovereignty in a Digital Environment', *Convergence*, 12:2, pp. 179–197.

O'Neill, B. (2007), 'Digital Audio Broadcasting in Canada: Technology and Policy in the Transition to Digital Radio', *Canadian Journal of Communication*, 32:1, pp. 71–90.

O'Neill, B. (2008), 'Digital Radio Policy in Canada: From Analog Replacement to Multimedia Convergence', *Journal of Radio & Audio Media*, 15:1, pp. 26 – 40.

O'Neill, B. (2009), 'DAB Eureka-147: a European vision for digital radio', *New Media Society*, 11: 1–2, pp. 261–278.

O'Neill, B. and M. Murphy (in press), 'Crossing Borders: The introduction and legislation for satellite radio in Canada', in L. Parks and J. Schwoch (eds.), *Down to Earth: Satellite Technologies, Industries and Cultures*, Piscataway, NJ: Rutgers University Press.

Patrick, A. S., A. Black and T. Whalen (1996), 'CBC Radio on the Internet: An Experiment in Convergence', *Canadian Journal of Communication*, 21:1.

Pizzi, S. (2004), 'Rethinking DAB North of the Border', *Radio World Online*, http://www.www.rwonline.com/article/4102. Accessed 1 June 2009.

Rathbun, E. A. (2000), 'Proceeding with digital radio', *Broadcasting & Cable*, 130: 17, p. 35.

Rudin, R. (2006), 'The Development of DAB Digital Radio in the UK: The Battle for Control of a New Technology in an Old Medium', *Convergence*, 12:2, pp. 163–178.

Stacey, W. (2007), 'Canada Eyes IBOC Additions to DAB', *Radio World Online*, http://www.rwonline.com/pages/s.0049/t.775.html. Accessed 1 June 2009.

Task Force on Digital Radio (1995), *The sound of the future: The Canadian vision*, Ottawa: Minister of Supply & Service.

WorldDMB (2009), 'Country Information for DAB, DAB+ and DMB', http://www.worlddab.org/country_information. Accessed 1 June 2009.

Chapter 8

Future Scenarios for the Radio Industry[1]

Marko Ala-Fossi

One of the crucial drivers of digital broadcasting across Europe has been the prospect and timing of analogue switch off (ASO). The target deadline of 2012 set by the European Commission for all Member-States to complete a conversion to digital terrestrial television broadcasting has created a firm timetable for the roll-out of digital terrestrial television across Europe, with a clear commitment to an all-digital future for the television industry. Radio broadcasting, for many of the reasons outlined in the foregoing, does not have this certainty. While some countries have begun to consider circumstances in which analogue radio switch-off might be feasible, the absence of a specific target date or an agreed vision has made the future of radio a matter of ongoing debate and speculation. For example, is switch-off for analogue radio realistic in the near to mid-future? What digital radio platforms are most likely to succeed and in what circumstances? To what extent is Internet delivery likely to impact on terrestrial transmission? This chapter presents the views of industry experts on possible future scenarios for radio as perceived from an industry standpoint. Based around the core research question of what the future of radio will look like in 2015, the analysis, while rooted in a fast-changing technology environment, outlines a number of general scenarios that map potential future configurations of radio delivery.

The Future Scenarios Study

Our research project to study the technological landscape and the future of radio was launched in 2005, when DAB had its ten year anniversary. The project was based on two earlier studies within the Digital Radio Cultures in Europe research group (DRACE): a comparative study of the development of DAB radio in the UK, Ireland, Denmark and Finland (Lax et al. 2008; Ala-Fossi and Jauert 2006), and a separate study of DAB in Canada (O'Neill 2007). The first stage of the present study was to map all the available – both existing and emerging – technologies for delivering audio (i.e. radio and radio-like services). The special characteristics, as well as the potential social and economic strengths and weaknesses of these technologies, were then analysed (cf. Chapter 2) and the data was collected in a database at the Intranet website of the project.

In addition to developing our own socio-economic understanding of the technological development of radio, we wanted to find out how the future of radio was perceived and understood by broadcasters, technology experts and other professionals working within the radio industry. In asking our interviewees to look forward to the next ten years of radio, we were necessarily inviting speculative responses, but the analysis of the interviews and their scenarios was intended to reveal which (if any) of the present different technological options they believed would succeed, and how they understood the current situation in which numerous options appeared to be in competition with each other.

The primary data of this study are 43 semi-structured expert interviews, which were carried out in Ireland (3), United Kingdom (13), Denmark (6), Finland (11) and Canada (10) between May 2005 and June 2006. Fourteen interviewees were public broadcasters, twelve commercial or private broadcasters, six regulators and five representatives of different economic or technological interest groups, three network or multiplex operators and three persons working for the media electronics industry. The interviewees were all in senior managerial positions in their own field and in some cases were leading members of industry groups. With only two exceptions, they were all interviewed in their native language. In addition, they were all asked the same basic set of questions which were adapted to the context of the interview. For example, the questions were not always asked with the same words and the questions concerning only Europe were dropped off in Canadian interviews. The interview questions were designed to explore how experts perceived the future of digital radio in terms of delivery, technology, socio-economic issues, displaced technologies and medium content. Most of the interviews were recorded during a personal meeting (39), but because of time constraints some (4) were made via e-mail or/and by phone.[2]

The method adopted for the study is that of qualitative content analysis of interview material within a grounded theory perspective. For practical reasons, each member of the research group analysed their own research interviews conducted in their native language. In addition, the understanding of the different national contexts was absolutely crucial for this analysis. The different nuances and emphases – or even opposing views between different nationalities – became more understandable in relation to the choices made in the past. Without a separate preliminary study, the different character and dimensions of the delivery technologies would have been lost behind the large number of acronyms and abbreviations.

Our preliminary analysis and understanding of the development of digital technologies, presented in Chapter 2, is based mainly on the social shaping of technology (SST) perspective (Mackay and Gillespie 1992; Winston 1998; MacKenzie and Wacjman 1999). We assume that technologies are always shaped by a combination of social, political and economic forces and processes. This is why their design and preferred forms of deployment will also match better with certain social, political or economic objectives and even exclude others. While a retrospective analysis of this social shaping of technology could readily be applied also to the development of digital radio technologies, this is beyond the remit of the present chapter. Nevertheless, some of the interviewees' responses clearly reflect the consequences of the social forces underlying the development of digital radio. However, we asked our interviewees to speculate on future developments starting from today's technology and that is why we have also applied the concepts of the diffusion of innovations theory (Rogers 2003). Despite their many fundamental differences, SST perspectives and the diffusion of innovations theory are not really antithetical and they both provide useful tools for analysing the development – and also appropriation or adoption of new technologies (Lievrouw 2006: 246–261).

Especially useful for our analysis has been the classification of perceived attributes of innovations, which according to diffusion theory explain most of the variance in the rate of adoption of innovations. *Relative advantage* is the degree to which an innovation is presumed to be better than the earlier system or idea. *Compatibility* is the degree to which an innovation is seen as consistent with the existing values, past experiences and the needs of the potential users. *Complexity* is the degree to which an innovation is perceived as relatively difficult to understand and use. *Observability* is the degree to which the results and possible benefits of an innovation are visible. *Trialability* is the degree to which an innovation can be experimented before adoption (Rogers 2003, pp. 219–265). For example, the low adoption of DAB in some countries, discussed elsewhere in this volume, might be interpreted as meaning that its relative advantage over FM was not considered truly significant, and its compatibility with other than nationwide public broadcasting systems was rather low.

Radio in 2015: future scenarios from national perspectives

Each and every research interview began with the same basic question: (Q1) *how do you think people will receive radio content in your country in 2015?* Depending on the answer, the respondent was then asked (Q2) *to explain his/her opinion and to evaluate the role of terrestrial analogue, terrestrial digital, satellite and Internet radio*, especially if they were not mentioned as parts of radio services in the first place. In addition, all the European interviewees were asked (Q3) *what will be the dominant way of delivering radio content in Europe?* Because both the interviews and their analysis have drawn on primarily national perspectives, the results are presented here following the same principle.

All the UK respondents believed that conventional terrestrial radio would remain significant in 2015, mainly because of its familiarity and portability. Most of them (8) also believed that terrestrial broadcasting in 2015 would be predominantly digital, with DAB as the most important platform, though there were likely to be additional digital platforms such as DRM. Those who did not believe explicitly that digital would dominate in ten years' time were the community broadcasters and, less definitely, smaller commercial broadcasters. Community broadcasters (2) thought that community radio should have a place on a digital platform but that DAB might not be the most appropriate. Although only two persons (both at BBC) believed that FM would have already been shut down, most UK respondents thought that FM would be a dying platform, appropriate mainly for new or alternative forms of radio, e.g. community radio. Two interviewees (both commercial broadcasters) also thought downloading/podcasting would become significant by 2015.

The Danish perspectives on radio in 2015 were in many ways very similar to those in the United Kingdom. Most of the respondents thought that FM radio would still exist in 2015 and even further on, but increasingly would be overtaken by digital transmission, primarily DAB and/or DRM. Also, in Denmark it is likely that community radio stations especially would continue to prefer to use the FM band for some time. However, some Danish respondents (2) were convinced that the Internet will become the dominant digital platform for radio within a few years – as soon as WLAN wireless coverage is extensive enough – and to which community radio and other niche channels would eventually migrate. People would then be able to receive all sorts of radio services with new kinds of multiplatform receivers. Other respondents were less convinced about the convergence of all radio media but some believed that the definition of radio might change over time. Radio will perhaps remain primarily an aural broadcast medium only in mobile reception, while in other contexts it may also offer other services. Although radio on-demand will co-exist with broadcasting, most radio listening will be linear and, as now, will take place while doing something else.

All except one of the Finnish respondents believed that terrestrial analogue FM radio would still be on air in 2015. However, their opinions about its importance were strongly divided. Half of the group thought that FM radio would still be the dominant radio platform, while the rest argued that the number of FM stations and their listeners would be decreasing by that time. In addition, they all thought that Finland would have some form of terrestrial digital radio broadcasting. Most believed in digital radio platform multiplicity and some were convinced that radio content will be received via multiple separate digital routes, including the Internet. Five respondents – including all YLE employees – thought that digital radio delivery in the DVB-T network will continue. However, most (9) respondents considered DVB-H as the most likely system for mobile and handheld digital radio, while DMB was mentioned only as a less likely option. None of the respondents supported the idea that DAB would be used as a digital radio system in Finland in the future. These opinions obviously reflect the idea that the shutdown of the Finnish DAB network in August 2005 was considered to be more or less irreversible (Ala-Fossi and Jauert 2006).

All the Irish respondents agreed that terrestrial analogue FM would still be popular in 2015 and that there would be some form of digital audio broadcasting in Ireland as well. However, the full range of views on the place of digital radio was evident. The respondent representing public service broadcasting thought that DAB or DRM would be strong, perhaps the primary platform of all. The regulator thought there would still be significant FM radio, while the commercial broadcasting respondent suggested there would be very little digital listening at all.

Some Canadian respondents (3) were somewhat reluctant to predict future reception platforms for radio and insisted that informative local content would provide the most important continuity. One respondent suggested that analogue broadcasting would become marginal in ten years, but most (6) were more conservative and believed that analogue terrestrial radio would remain pre-eminent in 2015. There was an understanding that digital radio would have an important place, though much less certainty about what form this would take. Public service broadcasters tended to have little faith in DAB and expressed greater interest in satellite and Internet radio. Most of the commercial broadcasters thought that subscription-based satellite radio would become important, though not dominant, in Canada. They had mixed views on terrestrial digital radio: some believed that DAB (with DRM and DMB) could become important, while others thought that it was already a failure and would remain marginal. IBOC was widely considered as a relatively weak technology with little added value, though some respondents believed that it could become a platform for radio in Canada, especially if it was able to succeed in the United States. [3]

There were, according to respondents in the study, many reasons to believe in the longevity of terrestrial analogue FM broadcasting. Perhaps the most obvious is the sheer number of existing analogue receivers and their ubiquity as integrated units in all sorts of devices. Not only are FM receivers and FM transmitters relatively cheap to buy and easy to use, but both the reception and audio quality in the service area are usually good enough for the majority of users. All this makes FM especially suitable for small-scale and local radio operations. Moreover, FM radio does not have the same external pressure for digitalization that analogue TV has because the international mobile telecommunications business has been more interested in frequencies vacated by the shutdown of analogue television than FM band frequencies. In some countries (Ireland, Canada), the decline of AM radio may even strengthen FM radio. Some respondents also argued that no existing digital broadcasting system could directly substitute for FM and at the same time provide significant benefits for the radio industry or listeners. In other words, none of the available digital systems alone has a significant relative advantage over FM, which is a proven, well known and relatively simple and cheap technology, highly compatible with the existing systems and socio-economic structures of radio broadcasting.

In general, European respondents thought that digital radio broadcasting in their home countries would be primarily, if not solely, terrestrial. Usually, the future was seen as an extension of the present situation (cf. Wright 2005: 92), so that most of the existing systems were expected to remain in place in 2015. However, it was interesting that even in the UK and Denmark, with heavy investments in DAB, many respondents were quite convinced that there is also a need for an additional digital radio system like DRM. Based on the responses, one of the main reasons for this would be the low compatibility of existing DAB system with local and community radio (Hallet 2005). In addition, DVB-T was seen as an option for the future only in Denmark and Finland. Finland was also an otherwise quite unique case, not only because there was so little faith in DAB – as well as in DRM – but also because very few people elsewhere saw DVB-H as a significant option for digital radio. The most obvious reasons for this emphasis – also identified by some of the respondents – are the Finnish national technology policy and the involvement of Nokia in the development of DVB-H (Lax et al. 2008). The lack of a shared vision about digital broadcasting was also reflected in the Irish interviews, where the only system seen as an option by more than one respondent was DRM. Finally, the Canadian perspectives about the future of digital radio reflected two major factors: the failure of the national DAB roll-out plan and a quite realistic acceptance of the influence that developments in the US radio market will eventually have on Canada (O'Neill 2007).

Practically all the European respondents agreed that satellite radio would not become important in their home countries – although some of them suspected that by 2015 there might be some sort of European satellite radio available. Europe was seen as a very difficult market for satellite radio because the continent is divided into 'tribal societies' not only by the number of national barriers, but also cultural and linguistic ones too. In addition, in Europe there are already nationwide and ad-free radio services in practically every national market (unlike the US, where satellite radio has had some success), which means that the demand for subscription satellite radio might be too low to make the necessary large investments profitable. There were also doubts about satellite radio reception as reception is usually not good indoors and the elevation angle might be too low in northern Europe. A European satellite radio would probably need an extensive network of terrestrial filler transmitters, and Europe-wide frequency clearance for such a network might be hard, if not impossible, to obtain. Satellite radio was thus seen as a highly complex system with very low compatibility and low relative advantage in European radio markets. In Canada, where both the US satellite radio systems – Sirius and XM –had just entered into the national market (O'Neill 2006: 193), respondents had a different angle. Although nobody expected satellite radio to take over the Canadian radio market, it was seen as a proven system with clear market potential and likely impact on the broadcast landscape. The public service broadcaster CBC had already entered into a partnership with Sirius Radio, partly because the system, with three geosynchronous satellites, was so compatible with its needs: Sirius would provide CBC with full nationwide coverage, including the northernmost parts of Canada.

Most respondents thought that Internet radio was already an important and growing form of radio delivery, especially for specialized and small scale services and in fixed domestic reception via broadband and WLAN. In the most radical visions, Internet radio would also be available 'everywhere' via wireless in 2015, though some technology experts characterized the future of Internet radio at its best as 'portable' but not truly mobile (e.g. for in-car reception). As mentioned earlier, the strongest believers in the development of Internet radio were among the Danish and the Canadian respondents, while it was seen as most problematic among the Finns, mostly because of copyright issues. Again, this difference reflects the national contexts: unlike in the other countries, a longstanding dispute over music royalty payments had in effect frozen the development of Internet radio in Finland (Ala-Fossi and Jauert 2006: 72–73).[4] In general, based on our study, it seemed that while most respondents had a comprehensive understanding of the relevant national market, they were not always fully aware of developments in other countries and so did not have in-depth knowledge of all the available options for delivering radio (for example DRM).

No single platform or technology was raised above the others by the European respondents: there was no clear consensus about the dominant European way of delivering radio in 2015. However, some common trends were evident. First of all, only a few people altogether thought that DAB still had any chance of becoming the dominant platform in Europe. Even in the UK and Denmark, some respondents argued that DAB was already an outdated system which would be supplemented or even replaced by DRM. [Note: interviews preceded the release of the DAB+ standard in February 2007, though its development was well known]. In general, DAB was seen as remaining an important platform for those countries where it already existed, but other technologies might become more popular in other European countries. Thus, the majority of respondents agreed that there would be several parallel distribution platforms for radio in Europe, depending on which systems are considered appropriate in each country, but also depending more generally on developments in the purposes of the EU radio operations and consumer preferences. While the Irish respondents did not have a common view on which was the best alternative to DAB, the majority of the Finns saw DMB and DVB-H as the main rival options for the further digitalization of radio delivery in Europe. Finally, although the continuing existence of FM radio in Europe was seen as clearly likely by most, only very few respondents argued that FM would still be the dominant European radio delivery system in 2015.

Spotting the dinosaurs: how to identify the winners from future losers

After outlining their views on the future of radio, each respondent was asked (Q4) *why they thought so and to explain their arguments by exploring the following perspectives: regulation, ownership, market penetration, economic issues, production practices, geographic coverage area, functionality and user practices,* reflecting the same set of eight different perspectives in the preliminary socio-economic study on the delivery technologies (cf. Chapter 2). However, as they were not strictly defined during the interviews, the perspectives worked here more as an inspiration than an instruction for the interviewees. In addition, the respondents were asked to (Q5) *name technologies that will not be dominant in the future, but would be still in use.*

Regulation

There was a wide variety of opinion about the role and impact of regulation among the respondents. Some claimed that regulation had very limited effects, while others argued that the lack of regulation or its inappropriateness could be decisive in digital radio development. Respondents, especially from the United Kingdom and Ireland, saw national regulation in the form of incentives as an important element in the relative success of DAB in the UK. The importance of economic incentives on technological choices in commercial broadcasting was also recognised elsewhere (cf. Ala-Fossi and Jauert 2006: 78). On the other hand, it was widely felt that national differences in government regulation and frequency administration would contribute to a fragmentary European situation for digital radio delivery. Respondents in Finland and Canada saw their national regulation system as more 'market-driven' than, for example, in Denmark. While most Finns considered regulation as a highly national issue in the future, some Danes argued that national legislation would play a steadily decreasing role and that EU media regulation would have a much more central role than previously.

In addition, some respondents also considered technology policy both on the national and European levels as a form of regulation.

Ownership

Mainly the Finnish respondents saw the ownership perspective as important. Two different dimensions were identified: the ownership and copyright of delivery technologies themselves, and the ownership and copyright of the media content delivered. Two respondents working in regulatory agencies stated that only open, non-proprietary technologies – like DVB-H – could have wider success in the long run. Neither markets nor governments would tolerate any tendencies towards corporate monopolization of delivery technologies. In terms of content, in addition to the problems with Internet radio in Finland, there was also an on-going dispute about copyright payments and music royalties for simulcasting in the DVB-H network (Ala-Fossi and Jauert 2006: 72–73). This probably explains why the Finns considered content ownership and copyright dimensions very problematic in particular: some also thought that expensive content copyright payments could effectively prevent the use of certain radio delivery technologies.

Market penetration

The existing levels of market penetration were widely considered as one of the major strengths of FM, but also as favourable for DAB. In those countries where a DAB infrastructure was already in place, the cost of installing a different infrastructure for an alternative technology would be prohibitive. This would also help the introduction of DMB, which could use the DAB infrastructure. On the other hand, where DAB had a high market penetration, any introduction of system updates such as the new audio coding systems of DAB+ would only be very gradual. However, the market penetration of DAB in the United Kingdom and across Europe in general was considered by some Irish and Finnish respondents to be rather low in absolute terms, despite the extensive marketing in those countries where it was regarded as relatively successful. Some Canadian respondents argued that if a technology achieved a high enough penetration in the larger US markets, it would mean that the same standard would later also be adopted by Canada, not least because large manufacturers will offer the same products for both markets.

Economic issues

Many of the problems with DAB were understood by the respondents as principally economic issues. DAB transmission was widely considered as expensive and uneconomic, especially for commercial and community stations when the costs were compared to analogue transmission or to alternative digital broadcasting systems. This relative disadvantage of DAB from an economic perspective is caused more by the design of its implementation (multiplexes) and basic architecture (audio coding system) than by multiplex operator pricing policy. At the same time, some respondents argued that an alternative multimedia system network with a more efficient audio coding such as DAB+ would provide not only higher capacity, but it would also be more economic, with better cost efficiency per bit for the content operators. New services were seen as potential sources of new income for broadcasters and multiplex operators and, for example, some Finnish respondents stated that DVB-H was attractive because unlike DAB it offered the opportunity to identify the users and sell both content and services directly to them. From the economic perspective, several respondents saw Internet radio and satellite radio as expensive solutions. Satellite radio would need huge investment, while Internet radio required more money to cover royalties and higher delivery costs for additional listeners.

Coverage

DAB was also perceived as having problems with coverage. In the United Kingdom and Denmark it had become obvious that community radio and small commercial stations were not suited to the existing coverage patterns of DAB multiplexes (cf. Corominas et al. 2006). The Irish respondents agreed that DAB coverage would not be compatible with the needs of local radio stations in Ireland, and this was one of the main reasons why DAB had not secured widespread support in that country. In Finland in particular, the larger private broadcasters were not interested in paying for coverage in the sparsely inhabited northern areas. Similarly, the cost of building and running a DAB network with comprehensive coverage would also be relatively high in Canada. In addition, the lack of robust DAB indoor reception was also acknowledged as a problem by interviewees both in the UK and Denmark. In contrast, FM radio was widely seen as well suited for community radio and Internet radio was considered a good way to extend its coverage where needed. In Europe, terrestrial delivery was also thought to have significant benefits over satellite: the ability to provide local programming, the ease of producing programming for audiences in a single nation, and ease of reception and consumption. Meanwhile in Canada, the extensive satellite service was seen as a good way to produce truly nationwide coverage of radio services.

User practices

Most Canadian respondents thought that user practices were important determining factors, and that those adopted by younger listeners in particular would be the most successful in long run. Some also believed that in-car reception would be a more important factor in North

America than, for example, in Japan, where more people commute by train. In comparison, fewer European respondents saw user practices as important. For example, only one person in the UK and two in Finland thought that the consumers would be key in deciding which platform would succeed. Furthermore, it was argued that DAB had the advantage that it is familiar to users as a digital form of 'conventional' radio. However, this kind of compatibility with analogue radio was not necessarily intentional but was more a result of the lack of development of DAB's multimedia capabilities; for example, in Denmark, DR marketed DAB by emphasizing the diversity of services instead of the new dimensions of the radio experience. Some Finnish respondents believed that the future users of digital radio would expect to have their receivers integrated into a cellphone-like, handheld personal media device. There were also totally opposite views: it was also thought unlikely that one application would cover all user practices – especially if it required an expensive terminal and a subscription.

Functionality

The technical functionality of a particular digital radio system was generally not seen as a very important factor. Some respondents even argued that everything else mattered more than the system's technical performance. Even very good functionality could not guarantee success but, on the other hand, insufficient performance or the lack of certain characteristics like good mobile reception could be a burden. In the case of DAB, some Finns saw its outdated audio coding and the lack of real multimedia capacity as fatal. At the same time, some respondents from the United Kingdom saw DAB as a proven system, even the *only* proven system, which could offer additional functionality, although this argument failed to acknowledge the recognition of DAB's poor reception in certain circumstances (e.g. indoors). As mentioned earlier, satellite radio's indoor reception was also questioned. DRM was thought to have good functionality as a modern system with better audio coding than DAB.

Production practices

Of all the possible factors for the future development of radio, production practices were considered to be among the least important. Some respondents even argued that this would not matter at all, although they were mentioned in occasional references to podcasting and Internet radio as alternatives to broadcast radio.

Other important technologies

Among the European respondents, DRM was the favoured option as an important but not necessarily dominant radio delivery system in the future. In the UK and Denmark there were also some who thought that DRM might not just exist alongside DAB, but perhaps even replace it. Almost all UK interviewees and some Danes also anticipated that in the near future electronic chips in radio receivers would be multi-platform (DAB/DRM/FM),[5] while some Finns suggested that software radio would provide the solution to platform multiplicity (Sabel 2007). Another radio platform widely accepted as important but non-dominant was the Internet, with all IP-based solutions like Internet radio and podcasting – and some Danish and Canadian respondents thought that the Internet would most likely become the *dominant* radio platform. DAB, DMB, DVB-H and mobile phone networks were also mentioned occasionally, but nobody suggested that satellite radio or IBOC would become even secondary platforms for radio delivery in Europe.

Screening the imagined dial(s): what will be on the radio(s) in 2015?

The two final questions of the interview structure were not about radio delivery or radio technologies as such, but instead the respondents were asked (Q6) *what they thought the content of radio media would be like in 2015 and what listeners or users would be doing with it.* If necessary, they were also asked to (Q7) *explain why they thought the content of radio in 2015 would be like that.* Some of the respondents argued that they were not experts in these issues, but they were willing to speculate about the future trends of radio and audio content.

In the responses, there were two main trends evident in responses from all countries. On the one hand, radio in the future was expected to offer much more personalised and specialised content – also with multimedia elements – which would then be actively selected by the users for listening whenever they found it convenient. This sort of non-linear use of radio, or on-demand radio, would be free from the strict time schedules and the linear flow of normal broadcast radio. On the other hand, several respondents were convinced that most radio content and its use in the future would be very similar to its current form – some even suggested that, thanks to new channels, there would also be a renaissance of the most traditional genres and forms of radio content. Some respondents however were rather pessimistic about the future of traditional and information-based, real-time linear radio programme services, but in most cases the mutual co-existence of these two developments was seen as possible or even very likely (cf. Scannell 2005).

Thus, it was believed that traditional broadcast radio, which is a very time-based, linear medium where you have to be tuned in to a certain channel at certain time and follow the schedule designed by the broadcaster to be able to get the content you want, would partly become a thing of the past. Instead, one of the common characteristics of newly emerging digital platforms for radio was assumed to be higher listener and user sovereignty, a result both of the increasing number of channels (satellite and terrestrial digital radio as well as the Internet) and of non-linear content delivery and consumption (downloading, podcasting and listening of recorded audio files). Besides circumventing schedules, listening to downloaded, recorded programmes will also help overcome poor signal reception, especially when on the move. Some respondents, emphasizing the importance of Internet, argued that future media users will not accept that they cannot use radio and television in the same way that they use newspapers and the Internet – they will not wait till a certain time of day to listen to the news or to jazz, as they become used to consuming media whenever and wherever they wish (Ofcom 2004). This was seen as a challenge to broadcasters who are used to designing their production with a certain programme grid and for only one delivery channel.

There were quite different views about the importance of multimedia aspects of digital radio delivery. Three Finnish respondents suggested that radio lacked visuality and that this would be an important reason for overall digitalization of radio content delivery, while two others thought that the most important multimedia characteristic would be direct sales of content and services to the users. Some UK respondents thought that the most likely multimedia features of digital radio would be downloading music and radio programmes for later listening, while one Irish respondent believed that digital radio's electronic programme guide would be reasonably attractive to listeners. Some respondents with a more technical background thought also that radio would find a totally new kind of interactivity with users which would go beyond the current phone-in, SMS, WAP and Internet applications. Others, however, were sceptical of the benefits of digital radio multimedia and thought that the Internet or mobile phone systems would be better for downloading data than digital radio broadcasting.

It was also widely thought that the increasing number of channels available would also lead to increasing specialisation of radio content – just as in satellite radio – so that there would be special themed channels for every genre of music, as well as more channels with no music at all. Some Finnish respondents thought that in the future a growing amount of radio or audio content would be produced (or re-produced) by audiences, and one respondent suggested that Internet radio delivery could also be used for non-edited, linear, real-time audio flow from some live events.[6]

Perhaps it is not surprising that among the respondents from every country there was a strong – although not unanimous – belief in linear broadcast radio programming with real time listening; in other words a belief that the traditional strengths of radio would be retained. Mobility, easy access, real-time broadcasts, localism and integration with the community, personalities, entertainment, as well as different types of traditional journalistic and artistic audio programming were all considered important. This kind of broadcast flow radio consumption was described as 'passive' radio listening or 'lean-back' radio. It was argued that in 2015 there would continue to be a demand for broadly similar radio content as now and familiar radio-type services that do not require too much from the user or the listener. In addition, most Danish respondents and many Finns thought that professionally produced radio content even in its most traditional forms will have its place in the future. However, some Danish respondents suggested that radio drama and documentaries, for example, have so far been severely handicapped by broadcast scheduling and their possible renaissance would be a consequence of the growth of on-demand radio. They also thought that even though the content of radio overall would be the much same as it is now, there would be differences in which types of content were linked to which platforms.

Discussion: future scenarios for radio media

Most respondents clearly believed that some sort of digital terrestrial radio broadcasting would exist in their home countries in 2015. DAB was seen as a strong option mainly in Britain and Denmark, but even there it was expected to have supplementary digital platforms like DRM and DMB. In Finland, DVB-H was considered as the most likely option, while in Canada it was thought that there might also be IBOC alongside DAB and satellite radio. With few exceptions, most respondents believed that analogue FM broadcasting would remain significant until 2015, both in Europe and Canada. There were varying views about Internet-based radio and audio services everywhere, but the idea of successful satellite radio services in Europe was rejected by most European respondents.

The European respondents did not agree on any single dominant digital audio delivery platform. On the contrary, most of them thought that there would be distinct national solutions and multiple, co-existing digital systems for radio delivery; that is, several different digital futures for radio instead of one, with DAB seen as particularly important primarily in countries with already existing networks. Many respondents argued that the relative advantage of DAB for both the broadcasters and listeners was low, and its compatibility with the existing socio-economic structures of radio broadcasting was not satisfactory. World DMB has actually tried to address these same problems with the adoption of new audio codec (DAB+) and closer co-operation with multiplex-free DRM, which will provide an alternative way to replace FM (DRM+) (cf. Chapter 2).

The explanation offered for this is that the original basic design of DAB reflects the socio-political and economic structures of European radio broadcasting in the 1980s. At the time when DAB was created, in many European countries, private and local radio broadcasting was either non-existent or about to be introduced, and radio in general was still largely dominated by national public radio broadcasters, as discussed in Chapter 1. However, over the next few years the structures, competitive settings and power relations of radio broadcasting in Europe changed in a fundamental way – but all this was more or less neglected in the development of the DAB system (Vittet-Philippe and Crookes 1986: 8–24; Rudin 2006: 167). In other words, when DAB was finally introduced in 1995, it faced a very different sociopolitical environment than that in which it was originally designed (cf. Ala-Fossi and Stavitsky 2003).

Here lies also the reason why DRM is now expected to complement or to replace DAB – although it has less capacity for data and sound, it does not challenge the existing economic and social structures of radio broadcasting to the same extent as DAB. Using DRM, each broadcaster can keep their own transmitters and networks and continue operations on separate frequencies, as they do currently with analogue radio, instead of joining a shared multiplex – and this also makes the DRM coverage patterns more flexible. In this way, it is much more compatible with the diversity of current forms of analogue radio broadcasting in Europe than DAB. An unresolved question is whether DRM will be able to offer sufficient relative advantage over analogue FM, both for the broadcasters and for the listeners. In any case, what this proves is that at the time of the interviews, the original vision of a transition of radio in Europe and across the world to one single, superior digital system had already failed for good. It remains to be seen whether the new versions of DAB (DAB+, DAB-IP and DMB) will be more successful.

It was interesting to see that the idea of the dual development of future radio content was so widely supported among the respondents in all five countries. However, this division into traditional, linear broadcast audio content with real time listening, and to a new, increasingly personalized type of content which is received on-demand and consumed off-schedule, in fact already exists: web-based radio and audio services in particular have recently expanded the territory of radio media and to some extent have overcome their earlier limitations. A prediction that assumes that both types of content and consumption will also exist in the near future is hardly radical: on the contrary, it could be described as merely realistic, or perhaps even a bit conservative. At the same time, there was no consensus about whether traditional, linear radio produced by broadcasters would still have a significant share in 2015, or whether the majority of users would prefer freedom from schedules and personalised content over real-time radio.

The development trends in adoption of radio delivery technologies and the development of radio content consumption are both very important and, based on our results, still highly uncertain issues. This is why they are perfect for scenario building: using these two variables, it is possible to create a scenario matrix to describe four different future scenarios for radio media. The X axis indicates the options in delivery technologies, while the Y axis describes the tendencies in radio and audio content consumption (Wright 2005: 98–99). Because all the respondents were asked questions about both the development and adoption of delivery technologies (Q1 – Q5) and about radio content and its consumption (Q6 – Q7), it is also possible to estimate how closely they could be identified with these scenarios (see Figure 1). The location of each respondent in the scenario matrix is defined here according to X and Y values between 0 and 2, allocated by members of the research group based on subjective evaluation of the interviewee's opinions, choices and arguments.

In order to make the matrix more accessible, it is useful to give a brief description of each scenario. 'Towers of Babel' is actually very similar to the present situation where free analogue broadcast delivery and real time consumption of live (or otherwise linear) audio content are still the most important ways to produce and use radio. Traditional broadcast operators are still offering the majority of the services. In addition, there are services using many competing digital technologies, but none of the technologies has a really dominant position anywhere and this is why digital radio remains small. In 'DAB DReaM', free broadcast delivery and real time consumption of live or linear audio content also remain the most important forms of radio. Again, traditional broadcasters deliver most of the services. However, besides strong analogue radio there is fast growing digital radio with a clearly dominant technology or a selection of complementary digital technologies (e.g. DAB/DRM/DMB). *'Digital Diversity'* means that digital radio has a different (dominant) design in different parts of the world and there can be very distinct national solutions. In addition, in most countries there are several competing digital systems for radio delivery. In some cases, linear radio-type audio is just a small fraction of the services available. Multimedia, subscription and on-demand audio services provided by producers other than broadcasters are gradually increasing and becoming at least as important as traditional broadcast audio. In *'Multimedia Market'*, digital radio is growing with one dominant technology or group of technologies (e.g. DVB-T/DVB-H), but the role of traditional broadcast radio operators is diminishing. Multimedia and on-demand audio is also offered by non-broadcasters and is becoming as significant as traditional broadcast audio.

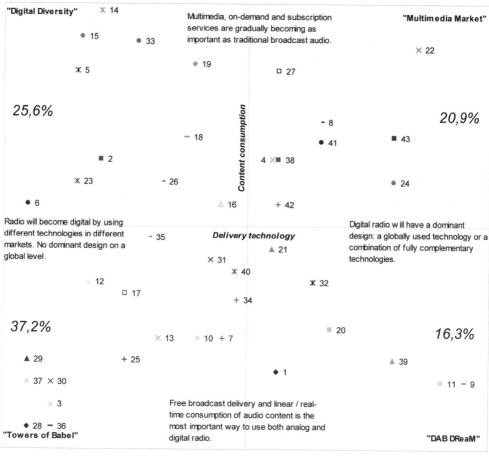

Figure 1: Division of the respondents between the four future scenarios for the radio media.

The respondents' locations on the scenario matrix appear to reflect their national contexts at least as much as their professional contexts. For example, most of the people supporting the '*DAB DReaM*' scenario were those other than broadcasters, from countries with existing DAB systems; broadcasters were also a small minority among the respondents supporting a future scenario like '*Multimedia Market*'. On the other hand, the supporters of a '*Towers of Babel*' scenario were mostly broadcasters, while a scenario like '*Digital Diversity*' was most commonly accepted, especially among public broadcasters.

These four scenarios, however, do not provide a fully comprehensive vision of the possible futures of radio media. The choices between the delivery technologies are more important the more exclusive they are, and these scenarios do not implicitly take into account any socially shaped development in reception systems. The present multiplicity of different delivery systems for radio has been shaped by complex social, political and economic processes. As a consequence, there is now a need for a new generation of receivers – or 'end user terminals'. A multiplatform receiver (e.g. AM/FM) is hardly a new invention and the latest development of reception devices has been very promising. In addition, a true '*Software Radio*' is no longer just science fiction (Sabel 2007). This sort of flexible, programmable radio receiver provides compatibility with a wide selection of delivery platforms. At least in the beginning, such receivers will also be complex and expensive devices – while the radio receiver as we have come to know it has been a cheap and simple to use device. Whatever the outcome of that development will be, the future of radio is always dependent on what people – the listeners – want to do with the content of the medium.

Notes

1. An earlier version of this chapter was published in part in the *Journal of Radio and Audio Media* (Ala-Fossi, Lax et al 2008).
2. There was never any intention to use the Delphi method (Linstone & Turoff 1975) for this study as a second round of interviews was not possible due to the limited resources of the project.
3. Experimental FM IBOC broadcasting was approved in Canada in October 2007.
4. An agreement over the Internet streaming music royalties in Finland was finally reached in June 2007.
5. This kind of multi-standard receivers became available in November 2006 (RadioScape 2006).
6. In Finland, for example some cities have started to deliver unedited live audio stream from their monthly city council meetings over the Internet (cf. http://www.espoo.fi). You may also listen to live stream from horse races (https://www.fintoto.fi/help/en/introduction/totoradio.htm) or live classical music concerts (http://www.e-concerthouse.com) over the Internet.

References

Ala-Fossi, M. and P. Jauert (2006), 'Nordic Radio in the Digital Age', in U. Carlsson (ed.), *Nordic Media Trends 9: Radio, TV & Internet in the Nordic Countries. Meeting the Challenges of New Media Technology*, Göteborg: Nordicom, pp. 65–87.

Ala-Fossi, M. and A. Stavitsky (2003), 'Understanding IBOC: Digital Technology for Analog Economics', *Journal of Radio Studies*, 10:1, pp. 63–79.

Ala-Fossi, M., S. Lax, B. O'Neill, P. Jauert and H. Shaw (2008), 'The Future of Radio is Still Digital – But Which One? Expert Perspectives and Future Scenarios for Radio Media in 2015', *Journal of Radio & Audio Media*, 15:1, pp. 4–25.

Corominas, M., M. Bonet, J. Guimerà and I. Fernández (2006), 'Digitalization and the Concept of "Local": The Case of Radio in Spain', *Journal of Radio Studies*, 13:1, pp. 116–128.

Lax, S., M. Ala-Fossi, P. Jauert and H. Shaw (2006), 'DAB: the future of radio? The development of digital radio in four European countries', *Media, Culture & Society*, 30:2, pp. 151–166.

Lievrouw, L. (2006), 'New Media Design and Development: Diffusion of Innovations v Social Shaping of Technology', in L. Lievrouw and S. Livingstone (eds.), *Handbook of New Media: Social Shaping and Social Consequences of ICTs. Student edition*, London: Sage, pp. 246–265.

Linstone, H.A. and M. Turoff (eds.) (1975), *The Delphi Method. Techniques and Applications*, Massachusetts: Addison-Wesley Publishing Company, http://is.njit.edu/pubs/delphibook/. Accessed 23 November 2005.

Mackay, H. And G. Gillespie (1992), 'Extending the Social Shaping of Technology Approach: Ideology and Appropriation', *Social Studies of Science*, 22:4, pp. 685–716.

MacKenzie, D. and J. Wacjman (eds.) (1999), *The Social Shaping of Technology; Second Edition*, Maidenhead and Philadelphia: Open University Press.

O'Neill, B. (2007), 'Digital Audio Broadcasting in Canada: Technology and Policy in the Transition to Digital Radio', *Canadian Journal of Communication*, 32:1, pp. 71–90.

O'Neill, B. (2006), 'CBC.ca. Broadcast Sovereignty in a Digital Environment', *Convergence: The International Journal of Research into New Media Technologies*, 12:2, pp. 179–197.

Ofcom (2004), *The iPod Generation. Devices and Desires of the Next Generation of Radio Listeners*, Prepared for Ofcom by the Knowledge Agency, http://www.ofcom.org.uk/research/radio/reports/ipod_gen/ipod.pdf. Accessed 15 September 2004.

RadioScape (2006), 'RadioScape now shipping multi-standard DAB/DRM modules to meet early market demand. First multi-standard consumer radios now on sale', Press release, 20 November 2006, http://www.radioscape.com/downloads/Press_Release/RS201106.pdf. Accessed 28 November 2006.

Rogers, E. (2003), *Diffusion of Innovations; Fifth Edition*, New York & London: Free Press.

Rudin, R. (2006), 'The Development of DAB Digital Radio in the UK. The Battle for Control of a New Technology in an Old Medium', *Convergence: The International Journal of Research into New Media Technologies*, 12:2, pp. 163–178.

Sabel, L. (2007), 'Software-defined radio – the solution for multi-standard multimedia in the mobile environment', *EBU Technical Review*, January 2007, http://www.ebu.ch/en/technical/trev/trev_309-radioscape.pdf. Accessed 10 February 2007.

Scannell, P. (2005), 'The Meaning of *Broad*casting in the Digital Era', in G. Lowe and P. Jauert (eds.), *Cultural Dilemmas in Public Service Broadcasting. RIPE@2005*, Göteborg: Nordicom, pp. 129–142.

Vittet-Philippe, P. and P. Crookes (1986), *Local Radio and Regional Development in Europe*, Manchester: The European Institute for the Media.

Winston, B. (1998), *Media Technology and Society – A History: From the Telegraph to the Internet*, London and New York: Routledge.

Wright, A. (2005), 'Using Scenarios to Challenge and Change Management Thinking', *Total Quality Management and Business Excellence*, 16:1, pp. 87–103.

Chapter 9

Community Radio In Transition: The Challenge of Digital Migration

Lawrie Hallett

Introduction

This chapter examines the evolving position of community radio in the context of digital migration and the increasing fragmentation of broadcast radio delivery. Summarising the transmission requirements of community radio stations, it considers these in relation to the use of digital broadcast radio platforms and goes on to examine some of the alternative delivery opportunities that now exist outside the confines of traditional broadcasting. Contextually, this chapter also considers the position of other types of radio broadcasting, and examines the potential influence that such broadcasters may exercise over the future development of the community radio sector.

Historically, the digitization discourse as it relates to broadcasting has typically been characterised by considerable optimism on the part of those developing the systems involved. Encouraged by such optimism, and by the promise of additional broadcasting capacity, politicians and regulators in many jurisdictions have driven forward the introduction of new digital broadcast radio transmission platforms. However, despite such official support, broadcasters and the public remain somewhat wary of investing in such technologies and, conversely, largely supportive of traditional FM broadcasting in particular.

Despite various digital radio transmission platforms now being either operational or nearing launch, it remains impossible to predict which option (or options), if any, will eventually emerge as accepted standards in the longer-term. This situation is further complicated by the increasing impacts of alternative non-broadcast content delivery methods. What is clear is that some digital radio broadcasting platforms are more flexible than others, and that some are best suited only to particular types or scales of radio broadcasting. Given the ubiquitous nature of FM transmissions, it also seems likely that in the majority of jurisdictions at least, this analogue platform's future remains secure for the foreseeable future. Therefore, this chapter argues for a nuanced approach to the delivery of community radio programming, selecting appropriate technologies on a case by case basis according to the specific coverage and content delivery requirements of particular stations.

Although in terms of broadcast frequency availability the advent of digital radio transmission platforms offers the potential to help reduce the imbalance between supply and demand, such developments do not herald a complete end to frequency scarcity. Frequencies allocated to broadcasting will remain finite and available bandwidth will by necessity remain limited. Inevitably competition for broadcasting rights will remain a barrier to entry for the foreseeable future. Assuming an ongoing requirement for access to the airwaves, the question for community radio broadcasters is how best to obtain usage rights to a greater percentage of the available frequency 'real-estate' than is presently the case. If the sector is to be successful in such endeavours, it needs to continue to build up its circle of friends. It will need to convince politicians and regulators of the strength of its case, something which may be easier said than done in the context of strong, well organised lobbying capacity available to competing public service broadcasters (PSB) and commercial operators.

Ask European politicians or regulators about community radio and they won't always know what you are talking about. Ask the same people about PSB or commercial radio and not only will they know what you are talking about, but almost certainly, they will also have opinions on the subject. The comparatively limited profile of community radio is in part due to the sector's relatively small-scale (both numerically, in terms of stations broadcasting, and in relation to the typical geographical coverage of such stations). However, it is also due to the fact that in most jurisdictions the sector is comparatively young, and therefore inevitably lacking in terms of track-record. It is a simple fact that relationships with politicians, regulators, funding bodies and partner organisations take a considerable length of time to establish and solidify.

Outside Europe, there are jurisdictions where community radio broadcasting is much more firmly established. A good example is that of Australia where not only has the community radio sector been in existence for many years, but it has also made some progress towards becoming part of a digital radio future. From its launch as far back as 1972, Australian community radio has expanded from an initial 24 stations to 351 in September 2008 (DBCDE 2009).

According to Digital Radio Australia (DRA), 'The Australian Government's Broadcasting Legislation Amendment (Digital Radio) Bill 2007 was passed in Parliament on 10 May 2007. It enables existing commercial, national and wide-coverage community radio broadcasters to provide digital radio services' (DRA 2008: 3). However, as the Community Broadcasting Association of Australia notes:

> The Federal Government has decided on a staged rollout of digital radio services, beginning in the six state capitals from 2009. As a result, the initial involvement of community broadcasters is limited to forty 'wide-coverage' stations. The Government hasn't committed to a full rollout of digital radio throughout Australia (CBAA 2009).

The key phrase above is that the Australian community radio stations initially involved in digital broadcasting will be limited to those with 'wide-coverage'. The CBAA goes on to note that analogue (AM and FM) broadcasting will continue 'for a considerable period' and observes that Australian Government policy realises that digital 'may never be a complete replacement' (CBAA 2009).

Clearly, the community radio sector in Australia has made some progress towards securing at least partial involvement in a digital future. Unfortunately, such success is not going to be easy to replicate within a European context. This is because the 'wide-coverage' community services referred to above are very much the minority within a European community radio context. Smaller-scale community radio services in Australia remain, like their cousins in Europe, largely excluded from current digital radio transmission platforms, which are simply not designed to accommodate a multiplicity of smaller stations each with its own distinctive coverage requirements.

Across Europe, community radio lacks the established support structures afforded to its longer-established PSB and commercial counterparts. Partly because of their greater longevity, but also because each can call on relatively predictable lines of income, both PSB operators and commercial stations have succeeded in developing strong lobbies to support and protect their operational interests. Whilst the European Broadcasting Union (EBU) is established to promote the interests of, and encourage collaborations between, the public service broadcasters of individual nation states, in parallel, the Association of European Radios (AER) works for the commercial sector.

By comparison, community radio has historically tended to have only limited centralised structures to call upon when promoting its cause. With a smaller financial base and considerable dependence upon volunteer support, the community radio sector has tended to be somewhat less well organised in terms of its campaigning and lobbying capacities. Because community radio has developed on a country-by-country basis, it is perhaps inevitable that some jurisdictions have developed stronger sector support bodies than others. For example, being well-established over many years, community stations in the Republic of Ireland and the Netherlands enjoy the support of relatively well-funded national organisations (CRAOL and OLON respectively). However, other countries, particularly those in Central and Eastern Europe, remain far less well developed in terms of their various lobbying capacities. Indeed, in some countries the debate remains not about how to regulate and encourage community radio, but rather about whether or not to permit the existence of the sector in any meaningful form at all.

In light of the above, at the European level it has proven more difficult to establish trans-national supporting bodies for community radio. Recently however, the situation seems to have been gradually improving. Bringing together practitioners, academics and other interested parties, the Community Media Forum Europe (CMFE) has emerged with a specific remit to lobby and campaign for community media (including radio) at the European level. In parallel, following a gap of several years, a European chapter of the World Association of Community Broadcasters (AMARC) has been re-established, to encourage greater collaboration and exchange between community radio stations across wider Europe. Nevertheless, there remains a considerable gulf between the lobbying capabilities of the community sector and those of larger broadcasters, and thus there remains considerable work for the sector to do in order for it to be able to prove its case and be treated similarly to its more established competitors.

Changing media and evolving platforms

As the senior electronic medium, broadcast radio has a long history. Evolving over time, radio has expanded both in terms of the number of stations broadcast and the nature of such stations. In a European context, following an early experimental period, most jurisdictions established public service broadcasting as the foundation of their radio provision. Later, legislative and regulatory frameworks were adapted and PSB providers found themselves the subject of commercial competition. More recently, European legislative and regulatory frameworks have gradually begun to change again, this time to accommodate community radio.

At the same time, however, broadcast media infrastructure is also changing. Internally, the medium is adapting to the emergence of digital radio broadcasting platforms, whilst externally, the effectiveness of so-called new media platforms is also creating opportunities and threats for broadcasters. The result of this combination of circumstances is that proponents of community radio seeking to establish and cement the sector as a robust and integral third-tier of radio broadcasting are doing so in an atmosphere of regulatory and technological uncertainty and flux.

Digital broadcasting platforms

Almost all currently available digital radio transmission platforms are based on the concept of a wide-band multiplex. In Europe, the longest established digital radio platform is Digital Audio Broadcasting (DAB) (Eureka-147 T-DAB) which was first put into regular use during the mid-1990s (ETSI 1997). This platform has since been joined by others, including DAB+ (ETSI 2007), Digital Multimedia Broadcasting (DMB) (ETSI 2009) and various flavours of Digital Video Broadcasting (DVB) such as DVB Terrestrial (DVB-T) (ETSI 2008) and DVB Handheld (DVB-H) (ETSI 2004). Over recent years, however, the boundaries of what constitutes broadcast radio have been changing, the result of wider digitisation of media. Today, radio stations are also transmitted via terrestrial and satellite digital broadcast platforms predominantly intended for television delivery.

In addition to the wide-band multiplex platforms summarised above, a second generation of digital radio transmission platforms is also under development. Amongst others, these include Digital Radio Mondiale (DRM) and DRM+, as well as HD Radio and FMeXtra, a system that enables the transmission of digital audio over FM sub-carriers (SCA). The key difference between these proposed platforms, all at varying stages of development, and their existent precursors is that instead of requiring separate, new, frequency bands within which to operate, each is able to operate within the traditional analogue (AM or FM) bands, in some cases even alongside existing analogue transmissions.

New media platforms

Alongside the development of platforms specifically designed for broadcasting purposes, new media technologies have also been impacting on the operation of broadcast radio. Not only do the Internet and the mobile phone provide alternative platforms for the delivery of linear radio in real time, but they also provide opportunities for the delivery of radio which is directly linked to other types of media content, and which include 'on-demand' elements that can be both time-shifted and non-linear, such as 'listen again' services and podcasts.

Although the use of such non-broadcast platforms can provide broadcasters with additional flexibility, they do not yet constitute a replacement for traditional broadcast platforms. There are several reasons why this is the case. For example, unlike one-to-many broadcasting platforms, both the Internet (as currently constituted) and the mobile phone networks are primarily designed as one-to-one communications platforms. In addition, for radio station operators, the economics of broadcasting are fundamentally very different from those associated with alternative (non-broadcast) platforms. Whereas broadcasters pay for range regardless of listenership, delivery via the Internet and mobile phone means paying on a per-listener basis regardless of where a particular listener might be in the world.

For the listener, at present, mobile phone and mobile Internet platforms lack universality, and tend towards end-user cost models which discourage the consumption of large amounts of data. However, it is clear that as the carrying capacity of mobile phone networks is enhanced, and as improved methods of mobile Internet delivery such as WiMax are implemented, this situation will change. In some jurisdictions 'all-you-can-eat' data tariffs are already becoming available (although connectivity and capacity both remain potential stumbling blocks to reliable portable operation). Nevertheless, convergence between broadcasting and communications platforms is already happening and as a result, after a long period of relative inertia, radio broadcasting is currently going through a period of ongoing change.

The changing nature of radio

As a nineteenth century invention, radio has survived as a viable technology for much longer than many other inventions of that period. The medium owes such success to an inherent flexibility and a proven ability to adapt. Broadcast radio is not simply just another commercially successful technology; rather it has become a social phenomenon, integral to our daily lives. Factors such as its immediacy and intimacy, as well as our ability to consume it whilst doing other things, all contribute to its success and enduring popularity. After almost one hundred years, AM radio may well be finally on the way out, but it survived and even thrived after the introduction of FM during the latter half of the twentieth century. Today radio is changing again, with FM currently moving sideways to accommodate the various flavours of digital delivery, which are gradually making inroads into the medium and pushing its evolution forward once more after many years as a relatively static medium.

In recent years, the emergence and rapid take-up of Internet-based delivery methods, already used by many broadcasters to supplement their traditional broadcasting platforms, has changed the debate over the future of radio. When digital broadcasting was first proposed, it was envisaged as a 'replacement technology' for linear real-time radio, superseding its AM and FM analogue forbears. Now, however, the situation is generally recognised as being more complex. Although the general public has not been overwhelmingly accepting of new broadcasting platforms such as DAB, it has embraced the consumption of non-linear 'radio' with alacrity, and there is no doubt that listeners now expect more than traditional real-time output from broadcasters. For example, BBC figures show that as early as October 2003 the corporation was delivering over 1.8 million hours of 'on-demand' content per month, a figure which had risen to over 4.0 million hours just one year later (BBC 2004). By January 2007 this figure had risen to over 7.1 million hours. During the same month over 14.5 million hours of 'live' BBC radio output were also consumed via the Internet (BBC 2007).

Specific figures relating to consumption of community radio content via the Internet are less easy to come by. However, a recent analysis of community stations in the United Kingdom, carried out in relation to this chapter, showed that of the 135 full-time community radio stations broadcasting as of May 2009, all but two had websites and no fewer than 120 (89%) were streaming their output in real time. Despite UK copyright restrictions which prevent community stations from providing 'listen again' services or podcasts that include copyrighted music content, 47 of these stations (35%) also provide some of their output either in podcast form or as streamed 'listen again' output. These figures clearly demonstrate that, despite the costs involved and the legal limitations placed upon certain aspects of Internet programme delivery, at least in a UK context, community radio stations are committed to the use of new media alongside traditional analogue broadcasting.

Community radio requirements

As a distinct, separate 'third tier' of radio broadcasting, there are some fundamental differences between community radio and its PSB and commercial counterparts. Such differences relate both to organisational structures and processes, as well as to operational objectives. Fundamentally, why community radio stations exist and what they try to achieve is different from other broadcasters. Such stations are distinctive both in terms of their inputs and their outputs.

There are some underlying commonalities which define community radio, such as operation on a not-for-profit basis, a commitment to accountability and to the involvement of members of the target community in the operation and management of the service concerned. However, a key feature of the sector as a whole lies in its diversity, each station is inevitably 'shaped by its environment and the distinct culture, history and reality of the community it serves' (Buckley et al. 2008: 207). Put another way, there is no such thing as a typical community radio service. Although many such stations serve clearly defined communities within a specific geographical area, others serve a specific community of interest, such as a particular ethnic minority or other demographic grouping.

Whilst geographical community stations may often have very limited but precise coverage requirements, because they seek to serve a specific sub-set of the community stations serving communities of interest will often have wider coverage requirements. So, in the context of digital broadcasting, what is it that community radio broadcasters expect from a transmission platform? Some of the key requirements are briefly summarised below:

- **Cost:** Although all broadcasters might complain about the state of their finances, budgets are typically particularly tight for community broadcasters that tend to depend heavily on volunteer support and unpredictable income streams. As a result, capital and recurrent (operational) costs are of particular concern to such broadcasters.
- **Complexity:** Community broadcasters require transmission infrastructure which is simple and well understood. The more complex the equipment, the more expensive it is likely to be as a result. Operational and maintenance costs are also proportionally related to complexity.
- **Coverage:** Community radio stations have coverage requirements which can be quite different from those of other broadcasters operating in the same area. Such differences may not only be in terms of scale, but also in terms of the specifics of the geographical locations to be served.
- **Universality:** Community radio stations, just like their PSB and commercial counterparts, require transmission infrastructure which reaches as much of their target listenership as possible. Broadcasters therefore look towards technologies which are used by as close to 100% of the population as possible.
- **Independence:** Again, often for reasons of cost, but in some jurisdictions also sometimes for political reasons and concerns over editorial independence, community broadcasters will typically seek to own and operate their transmission infrastructure independently.
- **Stability:** Like other businesses, community radio stations make investments for the long-term. They are therefore looking to invest in transmission systems with a predictable life-span.
- **Ease of maintenance:** On a practical note, primarily for reasons of cost, community radio stations often bring as much as possible of their operational maintenance 'in house'. Their ability do this is of course dependent upon the complexity of the transmission system used.

Analogue versus Digital

Focusing on the particular operational requirements of community radio broadcasters, as summarised in the preceding section, the following table provides an outline comparison of the relative merits of traditional analogue and current digital transmission platforms:

Table 1: Comparison of Analogue and Current Digital Broadcasting Platforms.

Aspect		Analogue (AM and FM)	Wide-band Digital (DAB etc.)
Costs:	Infrastructure / capital:	Low	High
	Operational /recurrent:	Low	High
Complexity:		Low	High
Coverage Flexibility:		High	Low
Universality / Penetration		High	Low
Independence:		High	Low
Stability / Permanence:		High	Low
Ease of Maintenance:		High	Low

The content of the preceding table clearly suggests that, for the present at least, when compared to their digital counterparts, analogue transmission platforms offer considerable advantages to community radio stations. A more detailed analysis, as set out below, explains why this is predominantly the case.

Analogue benefits

Costs (Infrastructure / capital): Analogue FM stereo transmission systems are relatively inexpensive to purchase and install. The technology involved is mature and well understood, it is also inherently simple and flexible – all of these factors contributing to its high degree of cost-effectiveness. In addition, economies of scale are achieved because the technology is internationally accepted with minimal variation all around the world. A further advantage of FM is that its transmitter installations often only require a small amount of equipment space from which to operate. Subject to terrain limitations, coverage of a medium-sized city (of say, 250,000 people) can often be achieved with a small stand-alone transmitter (of say, 100 to 500 Watts radiated power) operating from the top of a domestic residential tower-block.

(It should be noted that for a given coverage requirement, analogue AM transmission facilities tend to be considerably more expensive than their FM counterparts. This is primarily because AM antenna systems are inherently large structures. Not only do these need to occupy large areas of land but, because their AM transmissions can easily cause interference to other electronic equipment, they also need to be located away from housing or industrial buildings.)

There are various infrastructure cost premiums associated with current multiplexed-based digital broadcasting platforms. Fundamentally, they are inherently more complex than their analogue counterparts, adding additional steps to the transmission pathway in the form of multiplexers, and requiring more complex test and measurement. In addition, there are also further costs, which can be attributed to the relative immaturity of the technologies involved and to their less than universal roll-out, which prevent the achievement of similar economies of scale to those that are available when using FM. Finally, because of the fact that higher (Band III / L-Band) frequencies are used, the reality is that additional transmitter sites are often required to achieve particular geographical coverage, particularly within buildings.

Costs (Operational / recurrent): The recurrent costs associated with analogue broadcasting can also be minimal. They typically include as a minimum: transmitter licensing costs; electricity supply costs; audio feed costs (line rental or radio link licensing costs); and maintenance costs. In addition, there may be further costs associated with site rental or purchase; with required third-party compliance testing; and, in harsher environmental conditions, with heating, cooling and air-conditioning. A small, stand-alone analogue transmission installation, such as that which is typically used by a community radio station, can almost always be operated perfectly well from an independent non-commercial transmitter site.

For digital transmission systems, a similar cost base-line exists. However, particularly for multiplex-based systems, it is almost always the case that larger, dedicated transmission sites are used, which typically cost considerably more to rent than, say, the top of a local tower-block. The reasons behind this include the fact that programme material feeds from multiple stations have to be assembled together (multiplexed) prior to transmission. In addition, because they employ linear amplifiers which are less efficient than the non-linear equivalents which can be used in FM systems, DAB transmitters of an equivalent output power consume more current. As a result, electricity consumption costs tend to be higher, especially because additional cooling is often also required, thus consuming even more power.

Complexity: Analogue transmission systems are based on the principles of long-established technologies, which are well understood and for which the required skills and knowledge-base are widely disseminated. Where the knowledge-base to operate a technology is not widely disseminated, competition for equipment provision and support remains limited, thus placing upward pressure on end user costs for equipment and maintenance costs.

Coverage Flexibility: In order to best serve their particular target communities, community radio services often have very specific coverage requirements which may not be the same as those of nearby PSB and commercial stations. Analogue coverage (particularly FM) is planned on an individual station basis, adopting radiated power levels and polar radiation distribution to suit the specific coverage of each broadcaster as far as possible within local frequency resource constraints. Mutliplex-based digital transmission systems such as DAB require that all stations comprising a particular multiplex share identical coverage.

There is no possibility of one station obtaining different coverage from the rest. Many community radio stations have smaller coverage requirements than their commercial and PSB counterparts. Sharing the same transmission infrastructure may result in community radio stations achieving better coverage than would otherwise be the case, conversely, such a situation may result in such stations paying for coverage which they simply do not require.

Universality / penetration: AM and FM analogue receivers are ubiquitous, not only in homes, factories and offices, but also in cars and portable equipment such as mobile phones and MP3 players etc. By comparison digital radio receivers are quite rare, especially outside the home. For community radio, indeed as for all broadcasters, this means that the 'cost per potential listener' is currently considerably greater for digital radio broadcasting than it is for analogue.

Independence: Analogue (AM and FM) transmission infrastructures can often be operated as entirely independent installations. As a result, it is possible to operate them outside control or influence from third parties such as site owners or technical service providers.

Multiplexed digital transmission platforms which 'bundle together' a number of individual programme streams at a control centre prior to their broadcast, are by comparison particularly susceptible to such external control. With such systems there exists a third party 'gatekeeper' in the form of the multiplex operator, who has direct control over the programming stream, including the power to terminate transmissions.

Stability: For any broadcaster, investing in emerging technologies typically carries a greater degree of risk than putting the same financial resources into infrastructure which is firmly established. Technologies are sometimes superseded, and under such circumstances there is no such thing as a risk free investment. However, as with any other business, community radio operators are obliged to minimise such risks.

Ease of Maintenance: Because of their relative simplicity, the maintenance of analogue transmission installations can often be carried out 'in-house', providing a further opportunity to make recurrent cost savings. This is especially the case where transmitters are operated away from third-party commercial transmitter sites, as these will often charge 'access fees' each time on-site maintenance is required. Because multiplex-based digital transmission equipment tends to be housed at such third party sites, and because of its additional complexity, overall operational and maintenance charges will tend to be higher.

Digital benefits

It would seem from the above analysis that, for the present at least, analogue broadcasting platforms, and in particular FM, continue to be the most suitable transmission methods for the majority of community radio stations. By comparison, current first-generation, wide-band multiplex digital radio platforms such as DAB do not seem to offer a great deal in the way of benefits which might encourage their take-up by community broadcasters.

The situation is not entirely one-sided however. Even at their current stage of development and implementation, existing multiplex-based digital platforms such as DAB do offer some potential benefits, for example in terms of programme related data carrying capacity and robustness of mobile reception. The bottom line is that many of the benefits which such technologies bring are simply not of any great relevance to community radio broadcasters. Such stations do not usually have multiple programme streams to broadcast, nor do they typically need a network of transmitters in order to achieve wide-area geographical coverage. Thus the key benefits of multiplex transmission, such as the ability to delivery multiple programme channels and the ability to construct a single frequency multiplex (SFN), are simply not perceived as relevant to the sector.

Although in some jurisdictions, platforms such as DAB have been in operation for approaching fifteen years, internationally the development of digital transmission platforms has yet to reach maturity. Over recent years, various 'second generation' digital transmission platforms have begun to be developed which do not employ a wide-band multiplex approach, and thus offer the potential to provide digital transmission infrastructures which more closely reflect the various features of traditional analogue broadcasting.

Internationally, perhaps the most successful of these systems is DRM.[1] Intended as a direct replacement for AM broadcasting on frequencies below 30 MHz, it is already in regular use by various international broadcasters including the BBC World Service and Deutsche Welle. Because of the physics behind radio propagation, the original DRM standard is not suitable for use on higher frequencies such as FM Band II (87.5 to 108 MHz), as presently used for FM (stereo) transmissions. Recognising this limitation, the DRM consortium has recently been working on a new version of its standard known as DRM+, which is designed for use on frequencies between 30 MHz and 174 MHz.[2]

The DRM+ narrow-band transmission system has been specifically developed so that not only can it make use of frequencies within the existing FM broadcast band, but it can also maintain broad compatibility with FM frequency planning criteria. According to the limited technical information which was in the public domain at the time of writing, DRM+ will operate within a 100 kHz bandwidth with a data carrying capacity of between 37 and 186 kbit/s. The system is designed to be capable of carrying between one and four services in each multiplex (for example two audio services each with an associated ancillary data service) or to transmit programming in 5:1 surround sound (DRM Consortium 2008). The ability to deliver high-quality programming within such limited data-rates is a product of developments in the wider field of audio compression. Using MPEG-4 HE-AAC v.2 (High Efficiency Advanced Audio Coding) means that DRM+ achieves considerably better spectral efficiency than was possible using MPEG-2 audio coding, which was arguably the best available when the original Eureka 147 DAB standard was being developed some twenty years ago.

In recent comments about the evaluation of the system (DRM 2009), the DRM consortium noted a number of points which will be of specific relevance to community radio operators.

To begin with, it describes the system as being 'a perfect solution for stations not able to join [DAB etc.] multiplexes, even in places where the FM band is full'. According to a representative of the French community radio organisation, the Syndicat National des Radios Libres (SNRL), David Blanc:

> DRM+ seems to be an excellent choice, offering over 100 kbps of usable bit rate, enabling CD audio quality, slideshow and other data to be broadcast from a simple privately-owned transmitter. We now recommend integrating DRM+ in all digital radio receivers ... (DRM 2009)

Although the DRM+ standard has yet to be finalised, prototype tests of the technology have been taking place in Germany and France over recent months. Repeating Table 1 (above), but now adding in second-generation digital platforms, it becomes apparent that these do indeed begin to offer at least some of the benefits associated with traditional analogue broadcasting.

As can be seen from the table below, second generation narrow-band digital radio platforms possess at least some of the intrinsic attributes of traditional analogue broadcasting. Further analysis, as set out below using DRM+ as the example technology, explores the potential benefits of such systems in more detail.

Table 2: Comparison of Analogue with Current and Potential (second generation) Digital Broadcasting Platforms.

Aspect		Analogue (AM & FM)	Narrow-band Digital (DRM+ etc.)	Wide-band Digital (DAB etc.)
Costs	Infrastructure / capital:	Low	Higher	High
	Operational /recurrent:	Low	Low	High
Complexity:		Low	Medium	High
Coverage Flexibility:		High	High	Low
Universality / Penetration:		High	Low	Low
Independence:		High	High	Low
Stability / Permanence:		High	Low	Low
Ease of Maintenance:		High	Medium	Low

Costs (Infrastructure / capital): Infrastructure costs for narrow-band digital systems will probably be marginally higher than for an equivalent analogue installation. All new technologies tend to be subject to a price premium as suppliers seek to recoup development costs and, specifically for broadcast systems such as DRM+, there are also royalty charges associated with the use of MPEG audio compression standards.

Costs (Operational / recurrent): Operational costs for narrow-band systems are likely to be broadly similar to those of equivalent analogue transmissions. However, according to the DRM website, by comparison with FM transmissions, 'it needs lower transmission power for the same coverage' (DRM 2009). If this does prove to be the case, then operational cost savings may result.

Complexity: The complexity of narrow-band transmission systems is slightly greater than that of their analogue equivalents. The transmitters themselves are broadly similar, but the modulation employed is more complex.

Coverage Flexibility: The coverage flexibility of narrow-band transmission systems will be similar to that of FM with the exception that, with some such systems, for example DRM+, it will also be possible to change the way in which the available data capacity is allocated, in the same way as wide-band DAB multiplexes are sometimes reconfigured. For example, a station which usually transmits a single high-quality stereo audio programme with associated data could, on a temporary basis and perhaps in relation to a particular event, choose to reconfigure its output to consist of two separate audio channels, each of a slightly lower audio fidelity (or perhaps broadcast in mono). One such channel could continue to carry standard station output, whilst the other could be used to broadcast material of specific relevance to the event being covered.

Universality / penetration: Currently, the greatest weakness of narrow-band systems such as DRM+ is their immaturity. To date, none of these systems have been fully launched and receivers are not yet on the market. Thus, whatever the potential benefits of such systems may be, they remain some years off at best.

Independence: The level of independence provided by narrow-band transmission systems will be similar to that of FM. Without the need for programming from a variety of stations to be multiplexed together prior to transmission, it will be possible for individual stations to own and operate their own transmission systems as stand-alone entities.

Stability: Because of their immaturity, it is not yet possible to make an informed judgement as to the future potential of narrow-band transmission systems. Assuming that development proceeds successfully and receivers begin to reach the market, there may however come a time when broadcasters have to make value judgements about possible investments in such technologies.

Ease of Maintenance: In many respects the maintenance of narrow-band digital radio platforms is likely to be broadly similar to that of analogue transmission systems. The fundamental difference is that there will be a greater need for computer hardware maintenance in connection with the modulation scheme used.

Despite the promise of second generation digital radio platforms, it is important to bear in mind that such technologies remain many years away from public implementation. There is no guarantee that any of the specific systems currently under development will eventually become part of the mainstream radio industry. However, the underlying trend would appear to be towards the emergence of platforms which are increasingly able to deliver both the benefits of digital and the traditional flexibility of analogue platforms.

Limits on growth

Taking the United Kingdom as an example, since 2004 broadcast radio licensing there has been the responsibility of Ofcom (the Office of Communications). As a new regulatory body, Ofcom has taken a relatively pro-active approach to the introduction of community radio, offering licences to some 200 such services by early summer 2009. However, despite its recent rapid growth, the sector still suffers from a serious shortage of available frequency resources. For its second round of community radio licensing, begun in late 2006 and due for completion in the first half of 2010, Ofcom made it clear that there were 'a number of areas in the country where it is unlikely that suitable frequency resources remain available for further community stations to be licensed on FM' (Ofcom 2006: 2). Given that the number of additional community stations licensed during the second round will almost certainly more than double the number of such services operating in the UK, it would seem inevitable that, should a third round of community radio licensing be implemented from late 2010 onwards, this problem will increasingly limit the growth of the UK's community radio sector. On the plus side however, the recent closure of some commercial stations has released a certain amount of extra spectrum for use by additional community radio services.

Regulatory attitudes

To continue using the UK example, there the sector is helped by the existence of a supportive regulator which, early on, publicly stated its belief in the need 'to encourage the development of a thriving community radio sector' (Ofcom 2004: 11). In a subsequent document, the UK regulator expanded upon this view, stating that radio should develop to include:

> A multitude of community services at a very local level, providing social gain, community involvement and training for every community that wants and can sustain such a service, wherever they are in the UK. (Ofcom 2005: 26)

In addition, the UK regulator also makes clear that it understands the principles which underpin community radio, including the provision of:

… programmes for special interest groups (including ethnic and religious communities) providing a sense of identity in local communities, with community involvement and participation, broadcasting community information and allowing for debate. (Ofcom 2007: 22)

Whilst Ofcom clearly supports the principle of community radio for all, it also recognises the practical limitations of current transmission policies in relation to such services, noting that 'DAB is not generally suitable for community radio, given its multiplexed nature and so alternatives may be required. These could be digital (e.g. DRM) or analogue' (Ofcom 2005: 6).

Looking to the future, the UK regulator does however recognise the need for additional community radio frequencies and suggests a possible long-term solution:

In time, it is possible that changes such as an end to simulcasting of existing radio services on analogue and digital platforms could free-up spectrum that will create more space for new community radio stations. (Ofcom 2005: 28)

Although Ofcom is arguably towards the more enlightened end of the spectrum when it comes to community radio regulation, there are other jurisdictions which take a far less pro-active approach to the sector. However, many of the issues addressed by Ofcom are of international relevance and their approach does at least point to the usefulness of having a supportive regulator. Despite such support, it remains to be seen how successful Ofcom is in delivering against its goal of community radio for all. However, there is little doubt that in jurisdictions without appropriate and sensibly applied legislation and regulation, the future of the sector will be more difficult to secure.

Conclusions

Digital delivery methods are already impacting on the activities of community radio broadcasters, but not in the way that might have been supposed a decade or so ago. In the United Kingdom at least, the majority of community stations already make use of web-based 'new media' digital delivery opportunities to supplement their traditional analogue broadcasting output. Conversely, there is currently no apparent appetite for digital radio transmission platforms. Instead, recognising the various benefits of FM, the community radio sector is lobbying for greater access to Band II spectrum if and when other PSB and commercial broadcasters are persuaded to give up simulcasting and switch their broadcasting output to digital platforms.

There is however an element of risk associated with this approach. Specifically, there remains no guarantee that digital migration will be implemented, and without it access to additional FM spectrum cannot be provided. On the other hand, should digital migration be achieved for the majority of radio stations then community broadcasters remaining on FM could find themselves in what has by then become an 'analogue backwater' which the majority of potential listeners do not explore.

Nevertheless, given the largely inappropriate nature of current digital radio broadcasting platforms for community radio services, it is difficult to envisage how else the sector might approach this issue. However, it should not be forgotten that the current limitations of digital radio broadcasting are, to a large extent, technology specific. Emerging second generation platforms have the potential to be more relevant to the needs of community broadcasters, assuming that they do eventually become an integral part of the radio broadcasting landscape.

In practical terms, the potential emergence of digital radio platforms suitable for use by independent small-scale community broadcasters and similarly sized commercial stations remains, at best, some years off. Whilst it would be prudent for community broadcasters not to dismiss the future potential of such opportunities, continuing to exploit technologies which provide immediate benefits has to remain the priority. The approach of utilising web-based digital delivery methods, accessible through computers and mobile devices, is already providing increased flexibility and the ability to reach out to community diasporas which are not within the coverage of traditional analogue broadcasts.

Community radio broadcasters are typically, both by nature and necessity, pragmatists, seeking to serve their target communities in the most efficient and cost effective ways possible. Digital radio platforms may not be suitable today and whilst they may just become so in future, by that time it may well be the case that other non-broadcast solutions will have begun to dominate what today we call radio. Alternatively, FM radio spectrum may gradually be digitized, carrying an increasing number of DRM+ (or similar) digital services as analogue FM broadcasts are replaced and gradually begin to fall into decline. One thing that is already certain, however, is that the days of single platform analogue broadcasting are already fading into the past.

Notes

1. In the USA, 'HD Radio' is currently the preferred terrestrial digital radio system. It is also being trialled elsewhere, particularly in Central and South America. However, several American short-wave broadcasters have opted to use DRM for their international transmissions.
2. The Extension of DRM to Frequencies up to 174 MHz: DRM+ (http://www.drm.org/drm-the-system/drm/).

References

BBC (2004), 'Millions Flock to BBC Radio Online', 23 November 2004, http://www.bbc.co.uk/print/pressoffice/pressreleases/stories/2004/11_november/23/radio_online.shtml. Accessed 15 June 2009.
BBC (2007), 'BBC radio websites set standard for weekly users', 8 March 2007, http://www.bbc.co.uk/pressoffice/pressreleases/stories/2007/03_march/08/stats.pdf. Accessed 15 June 2009.

Buckley, S., K. Duer, T. Mendel and S. Ó. Siochrú (2008), *Broadcasting, Voice, and Accountability*, Ann Arbor: University of Michigan Press.

CBAA (Community Broadcasting Association of Australia) (2009), 'Digital Basics', http://www.cbaa.org.au/content.php/498.html. Accessed 24th October 2009.

DBCDE (Department of Broadband, Communications and the Digital Economy (2009), 'Community Radio', 4 June 2009, http://www.dbcde.gov.au/radio/community_radio. Accessed 24th October 2009.

Digital Radio Australia (2008), 'Digital Radio In Australia', May 2008, http://www.digitalradioaustralia.com.au/files/uploaded/file/Digital_Radio/cra-digitalFAQs-May2008.pdf.

DRM Consortium (2008), 'DRM+ Digital Radio Mondiale PowerPoint Presentation', August 2008, http://www.drm.org/fileadmin/media/downloads/drmplus_presentation_v1.6.pdf. Accessed 15 June 2009.

ETSI (European Telecommunications Standards Institute) (1997), 'Radio broadcasting systems; Digital Audio Broadcasting (DAB) to mobile, portable and fixed receivers', ETS 300: 401.

ETSI (European Telecommunications Standards Institute) (2004), 'Digital Video Broadcasting (DVB); Transmission System for Handheld Terminals (DVB-H)', ETSI EN 302: 304 V1.1.1, November 2004.

ETSI (European Telecommunications Standards Institute) (2007), 'Digital Audio Broadcasting (DAB); Transport of Advanced Audio Coding (AAC) audio', ETSI TS 102: 653.

ETSI (European Telecommunications Standards Institute) (2008), 'Digital Video Broadcasting (DVB); Implementation Guidelines of DVD Terrestrial Services; Transmission Aspects', EN 300 744 V1.6.1.

ETSI (European Telecommunications Standards Institute) (2009), 'Digital Audio Broadcasting (DAB); DMB Video Service; User Application Specification', ETSI TS 102 428 V. 1.2.1.

Ofcom (2004), 'Radio – Preparing for the future (Phase 1 developing a new framework)', http://www.ofcom.org.uk/consult/condocs/radio_review/radio_review2/radio_review.pdf. Accessed 15 June 2009.

Ofcom (2005), 'Radio – Preparing for the Future (Phase 2: Implementing the Framework)', http://www.ofcom.org.uk/consult/condocs/radio_reviewp2/p2.pdf. Accessed 15 June 2009.

Ofcom (2006), 'Community Radio Licensing – Second Round Statement', http://www.ofcom.org.uk/radio/ifi/rbl/commun_radio/ tlproc/secondround.pdf. Accessed 27 March 2009.

Ofcom (2007), 'The Future of Radio The future of FM and AM services and the alignment of analogue and digital regulation', http://www.ofcom.org.uk/consult/condocs/futureradio/future.pdf. Accessed 27 March 2009.

Chapter 10

Two-way Radio: Audience Participation and Editorial Control in the Future

Lars Nyre & Marko Ala-Fossi

This book deals with the promise of digital radio. In the 2000s, we see clear tendencies towards a participatory turn in broadcast media, and this is often praised as one the greatest improvements brought along by the digitalization of mass media. Broadcasters now commonly use SMS-messaging, interactive websites and talent shows to involve audiences more strongly in the editorial process (Jenkins 2008; Enli 2007).

However, this turn towards direct audience participation involves a tension between the (free) participation by audience members, and the increased level of registration and control by broadcasters. Participants leave rich trails of information that can be used and misused by those controlling it.

This book systematically interrogates digital technologies; for example, their potential for strengthening community radio (chapter 9), and the potential of podcasting to make radio non-linear (chapter 11). This chapter analyses the role of audience registration in all the prevalent platforms for radio media around 2010. It gives a brief historical account of the traditional anonymity of analogue radio and shows how the era of 'non-registration' is coming to an end. We discuss central features of the new techniques of participation afforded by digital media from the perspective of broadcasters and Internet service providers, and it also discusses the accompanying terms of registration and access. In part I, we introduce these features in a historical context and catalogue the available options in 2009. In part II, we present a detailed inventory of participation and registration features in three media platforms that will be central in the future: digital broadcasting, broadband Internet and mobile telecommunication networks. We conclude that there is a systematic correlation between increasing room for participation by audiences and increasing level of detail in registration by providers.

Part I: Goodbye to anonymity

While radio was previously a one-way medium, spreading its messages indiscriminately to a great number of individuals, it is increasingly becoming a two-way medium, staying in close connection with its individual users on new technological platforms. It is important to remember that this emerging two-way communication implies technical registration of all communicators. For example, there is no way to send an e-mail without also sending a signal that can be traced back to the computer and the person who sent it. The implication is that the user is always dependent on the decisions of broadcasters and providers, and under their editorial control.

The story really begins with analogue broadcasting in the 1920s. On this platform it was impossible to register who was listening to sound broadcasts and, consequently, providers could not require payment from their users. The absolute anonymity of the listener was in a sense an anomaly, a communicative side effect of the broadcasting infrastructure. The financial problem was solved by introducing license-fees and advertising, and the idea of an anonymous audience was accepted as integral to the analogue broadcasting platform. On the institutional level, it also resulted in the development of audience research that monitored the current trends of public interest, rooting editorial policies in statistical contact with the national audience.

It is interesting to note that analogue broadcasting contained a great unintended freedom for citizens in that they could listen freely to whatever they wanted. You cannot subscribe to a political newspaper without the paper, your mailman and even your neighbours knowing about it, but you can listen to a political broadcast without revealing your political orientation to anyone. This was demonstrated well in Norway during World War II when the German invasion forces prohibited radio listening and confiscated receivers on a national scale. Despite their efforts, they were unable to stop the illegal listening to Allied radio services and the London programmes became a vital source of information for Norwegians (Dahl 1999).

In addition to providing absolute anonymity for the audience, traditional broadcasting has another striking feature: it does not encourage audience participation. Since the listeners had no means of staying in direct contact with the providers, it was no wonder that the editorial content was created without their contribution. The manifesto of public service broadcasting has been to inform, educate and entertain the dispersed public. The institution is based on an ideology of the professional journalist and editor who create high-quality content for the general public. Jauert and Lowe (2005) promote the 'Enlightenment Mission' of traditional broadcasting as a primary value for the future.

However, there is a complication which Jauert and Lowe are the first to point out. In the twenty-first century, we are at a turning point regarding the role of the audiences in their media behaviour. It is abundantly clear that audiences want to contribute to public communication and media debate. There are thousands and millions of people who regularly express themselves on the Internet. Facebook and YouTube demonstrate the range of techniques for participation that people can engage in. Audiences everywhere can write, film, record, speak, edit, design, manipulate and publish their material in a myriad of ways on a myriad of digital platforms.

There is also increased mobilization. Handheld devices like mobile phones, radio receivers, mp3-players, palm computers and lap-tops are well suited to accommodate new types of audience participation. The devices are intimately associated with flesh-and-blood individuals and are typically off-limits to most other persons (Bull 2000; Ling 2004; Katz and Aakhus 2002).

But most digital media do not afford the luxury of anonymous media consumption. There is a give-and-take logic at play in this relationship; the users get access to far more diverse services than before, but in addition, their behaviour while using the platform is registered in great detail by the providers. Providers can disclose the user's geographical location by geo-positioning systems; his social network from hacking into his address book on the mobile phone; the content of his work life from logging his computer movements, and so on. It seems clear that digital media contrast starkly with analogue broadcasting when it comes to both registration and participation features (Fernback and Papacharissi 2007).

Techniques of participation

Our comparative analysis consists of pointing out the relevant features of participation for the many different platforms of media communication. We will analyze the techniques of participation in broadcast media on a continuum, from very active participation to completely private consumption behaviour. We believe that the act of speaking is the most potent means of public participation (Tolson 2006) and all platforms will be evaluated according to their affordance of public speaking by audience members. The list is not comprehensive but points to the most relevant techniques in our context.

Direct publishing by the citizen
1. The users produce their own speech and audiovisual material and publish it without prior agreement with the providers of the service, for example, by uploading a home video on YouTube. This is the most active form of participation in our scheme.

Participation in an editorial setting
2. The users seek out geographical locations to speak with the editors, based on GPS-location services in combination with a broadcast programme. This is a very active form of participation, but still quite rare in the 2000s.
3. The users speak or sing on air, either live or on a recording made by producers or the users themselves. This is a traditional form of participation which is completely controlled by the editorial unit.
4. The users write messages that are displayed on the TV screen or read out loud on the air. This is under greater control of editorial units than live verbal performances since they can very easily be edited.

Private media consumption
5. The users can create personalised playlists, for example, in Media Player, and control the type of music they want to hear. The same is the case for LPs and 78 rpms picked out from a personal record collection.
6. The listeners can access background information on the Internet; e.g. about a radio personality. This gives added value to the music or radio experience.
7. The users can time-shift the programmes on advanced radio and television tuners; e.g. TIVO. This makes it easier to incorporate media consumption in their domestic life rather than strictly live programming.

8. The users can change stations at their leisure, from a menu of options. Again, this is a way of controlling what they hear, but it is less comprehensive than playlists since users cannot choose what the station contains but can only tune in or out.
9. The users can switch the equipment on and off, and this is the basic privilege that secures individual freedom for the users, as long as the registration procedures are also shut off.

In addition to these direct forms of participation with the medium, there are also back door channels which rely on other media used in parallel. Subscribers receive invoices in the mail, and audience members can write letters of complaint or praise and send them in the post; or can call the switchboard and voice their opinions; and send e-mails or post messages to forums. This study does not put great weight on these back door channels because they don't affect the programme content directly.

Terms of registration set by providers

Audience control is at the heart of the modern media industry (Beniger 1991). There have been subscription newspapers for hundreds of years, license fees for radio and television since the 1920s and subscription cable TV since the 1940s. On the Internet, there are credit card payments through banking systems like PayPal, Ogone or the like. Indeed, since credit cards are linked to a person's name, they can be used to discover other information, such as postal address, phone number, etc. Moreover, registration can occur through the very device that the person is being charged for using and the telephone system is a good example of this. Providers can charge users for the duration of use or the amounts of data transferred, and register the temporal characteristics of the users' habits.

We will analyze new media platforms according to the list of registration procedures below. Notice again that the list is not exhaustive although it includes the most typical forms of control (see also Ala-Fossi 2005a; 2005b; Ala-Fossi, Lax et al., 2007).

1. The provider can shut off the user's access (due to non-payment or censorship).
2. The provider can record the individual's location at any given time.
3. The provider can record the individual's long-term media consumption history on a given platform or service.
4. The provider can identify and track the terminal and user through the phone number, and record all conversations.
5. The provider can identify and track the terminal and user through the IP-number of computers and record all traffic.
6. The provider can require username and password in order to allow the user access a site or service.
7. The provider can require a subscription scheme with codes that are changed regularly and sent out on e-mail, etc.

In most cases, the receiving device can be used by anybody in the household or institution, and the provider cannot necessarily trace a flesh and blood person. The latter will involve personal passwords or even fingerprint registration. Computers in a business or state agency may be used by several people for job purposes only, and in such cases it may be difficult to point out the person who actually sent the message. The flesh and blood aspect of registration in digital media will become ever-more important because of its commercial potential. Mobile phones and other handheld devices display the opportunities clearly in that they are typically owned and used by only one person, at least in the Western world. The person programmes the terminal with his preferences for work and leisure, relating to everything from bass and treble for the audio reproduction to his Facebook profile. People's varied activities on handheld devices potentially allow the providers to monitor their entire 'media lives'.

Notice that there are of course also editorial tools of registration (Ytreberg 2004). A webmaster may allow nicknames in a forum as long as he knows the user's identity. A caller to a radio talk show may be allowed to speak anonymously on the condition that he is positively identified by the editors in advance. Such editorially sanctioned anonymity can also be illustrated by the long chain of SMS messages on TV, where the provider could easily have identified all 'texters' by name on air but chooses not to. These forms of editorial registration will not be central to the platform analyses of this chapter although they are clearly an important feature of participation in mass media.

Let us rephrase the basic dilemma: At the same time as handheld digital media provide a really good context for public participation, the providers can keep track of every move the user makes. Lawrence Lessig warns against the regulation of personal behaviour that the digital 'architectures of control' afford (Lessig 1999: 33). People who express themselves safely within the norms of conventional society may not experience problems, but people who promote politically incorrect ideas, or behave in a provocative fashion for artistic or opportunistic reasons, can easily be sanctioned by providers. The knowledge about these sanctioning mechanisms can lead to self-censoring; the more sensitive the topic is, the less likely it is that people will speak up in public to oppose or promote it. The only realistic approach is to acknowledge that participation in audiovisual media in the future crucially involves registration and personal identification of all users. The more aware the public is of this double-bind, the better mass communication will become.

Part II: Inventory of participation platforms

The remainder of this chapter consists of a detailed account of the techniques of participation and terms of registration for nine sub-platforms in the three dominant digital platforms: digital broadcasting, broadband Internet and telecom networks. Figure 1 shows the scope of the inventory in this chapter. There are three digital distribution systems, with nine selected media platforms.

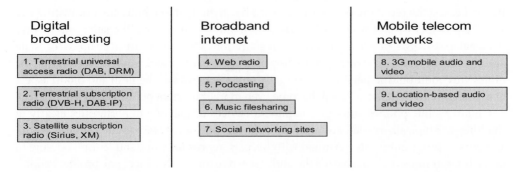

Figure 1: Inventory of participation platforms.

The empirical material for this account was generated by a comparative technology review inspired by medium theory (Meyrowitz 1994), and by systematic media histories such as those of Briggs and Burke (2004) and Winston (2005). We analyze the available techniques of participation and terms of registration in a series of nine digital media platforms, and we compare them in order to evaluate which platforms afford the best balance between participation and registration. For a detailed discussion of this approach to media studies, see Nyre (2008), and Nyre and Ala-Fossi (2008).

Digital broadcasting

The dominant speech functionality in digital broadcasting is journalism, that is, editorialized speech. Over the years, a very strong culture of content production has developed in radio and television, with professional norms for speech in various genres such as sports broadcasts, talk shows, documentaries and news. This eloquence gives broadcasting a very comfortable communicative presence for audiences. If you were ever to sit back and just receive, broadcasting would be your medium of choice because of the sheer aesthetic allure of its sounds (Crisell 1994; Hendy 2000).

In having such an important function, broadcast journalism rises above the communication that ordinary citizens can effect between themselves. Journalists work in a well-defined profession with trade unions and interest organizations; they possess complex expressive skills involving writing, camera work, styles of speaking and moving around, editing, checking sources, complying with ethical guidelines, etc. (Murdock 2005). It seems that digital broadcasting will continue to support this traditional, professional regime and not affect anything resembling a radical change in broadcasting institutions such as the BBC.

1. Terrestrial universal access radio
Digital audio broadcasting (DAB) was the first operational digital radio platform and was launched in the UK in the late 1990s (Lax, Shaw, et al. 2008). It was not only designed for

delivering sound, and in practice could include any sort of data like text and pictures – even video clips or web pages. One transmitter is able to broadcast several programmes at the same time on the same frequency, but this requires that programme services or channels are combined in a multiplex. DMB (Digital Multimedia Broadcasting) is an audiovisual update of the DAB system. It provides similarly good mobile reception, combined with new audio and video encoding standards, which makes it more versatile and efficient.

Regarding the techniques of participation among listeners, current DAB and DMB radio only allow three of the private consumption techniques: time-shifting the programmes on advanced tuners, changing stations on a menu, and switching on and off. This is because the signal stream only goes one way, with no return-channel for the listener. This feature also implies that there are no terms of access set by editorial units and providers. They cannot register the listeners in any way beyond registering the purchase of the listening equipment and collecting license fees from equipment owners.

Both these standards are based on the traditional broadcasting paradigm: it gives the listener free access to all services and content, and it also allows the listener to remain anonymous. This is why terrestrial DMB services in South Korea are financed only by advertising and the operators are allowed to simulcast their existing analogue terrestrial television and radio services. If this kind of anonymous digital reception were to dominate in the future, there would not really be any change in the radio medium from the perspective of our analysis.

We will mention two more digital terrestrial platforms with universal access that share the characteristics described above. DRM (Digital Radio Mondiale) also follows the traditional broadcast paradigm. It was developed to provide near-FM quality sound on AM and it also has capacity to broadcast additional data and text. In addition, it is designed to use the existing channel allocations. IBOC (In-Band, On Channel) or HD radio has a similar kind of approach. It uses the so-called sidebands on both sides of the analogue signal for the digital signal. This makes it possible to broadcast simultaneously a standard analogue signal plus one near CD-quality digital signal, and a small amount of additional data (Ala-Fossi and Stavitsky 2003: 67). In both cases, broadcasters can keep their own transmitters and networks and continue operations on frequencies that are well known to the listeners, instead of having to regain their positions on a single frequency network (like DAB multiplexes).

2. Terrestrial subscription radio

The pan-European Digital Video Broadcasting (DVB) project was launched in 1991 based on experiences with DAB (Lax, Shaw, et al., 2008). The latest branch of this family of standards is a special system for handheld devices called DVB-H. It primarily broadcasts programmes to everyone in the same way as traditional broadcasting but the user can also download on-demand type audio content that is not a strictly live transmission. In other words DVB-H allows the user to handle data packages at his leisure and this is a substantial difference from DAB.

Regarding the issue of registration, it is important to notice that DVB-H was designed to provide an IP-based platform for public communication. The IP-based system of DVB-H makes it possible to locate and register users and charge them for different services, which gives more possibilities for commercial use than DAB. This is why mobile phone manufacturers like Nokia, as well as telecom operators, are now pushing DVB-H to be the dominant standard for mobile digital broadcasting.

Channels can be broadcast free to air or they can be encrypted and available only via subscription. For example, the Nokia N77 supports such subscription channels and offers a limited degree of user management of such services. Subscription services can be available for fixed periods of time or for a particular event. DVB-H also supports interactive services like voting and participating in quizzes or surveys. Since DVB-H can be embedded in mobile phones, audience members can use the cellular connection to send information back to the service provider. This introduces a level of two-way interactivity that has not traditionally been available on radio and TV (except for the 'red button' that satellite and cable TV can provide, and which also uses a telephone feedback loop). With widespread adoption, these DVB-H services have great potential (Blandford 2007).

Regarding the balance of participation and registration, DVB-H only offers private consumption techniques of participation for citizens, although these are highly versatile. Listeners can purchase on-demand services which are actively selected; they can change stations on a menu (depending on the offer in their region); they can time-shift the broadcast programmes on advanced tuners, and they can turn their equipment on and off.

Along with these rich consumer opportunities, there is a much tighter control from the providers; or they at least have the technical possibility to exert it. The editorial units and providers control the listener's access by subscription codes that are changed regularly, and sent out on e-mail, etc. They can instantly shut off the user's access (due to non-payment or censorship); and they can record the (approximate) location of the listener's device at any given time based on the network characteristics of the IP-number. Again, it is clear that the increase in versatility for users is paid for with greater detail of registration by providers.

3. Satellite subscription radio

Geo-stationary satellites have a very high data capacity, at least downstream. Satellite radio can cover large areas like continents, and offer hundreds of options simultaneously to potentially millions of people. Satellite radio has been relatively successful in the USA, mostly because of its uninterrupted and crystal clear coast-to-coast coverage, with a variety of channels with little or no advertising. The market for satellite reception devices really only exists in the USA, where there are dozens of different models available. The satellite radio receiver resembles the TV receiver in many regards. There may be a recording functionality, programme pause and resume, and a graphical display of available services. An important difference is that there is no dish and the reception is therefore mobile. Especially for the car, there are devices with GPS functionality and radio reception, in addition to an mp3-player.

The techniques of participation for citizens are the same as for DVB-H and its relatives, except that it is not possible to download programmes and listen on-demand (this can only be done if the station also has an Internet service). The user can change stations on a menu; time-shift the programmes on advanced tuners, and turn the receiver on and off, and these are all types of participation that we call techniques for private consumption. On the other end of the platform, the editorial providers control the audience's access by tying subscription to codes that are renewed regularly on e-mail or in the post, and therefore they can shut off the user's access (due to non-payment or censorship). They can also identify the location of the device to a lesser or greater degree based on the density of the transmission stations (for example, on the level of city or suburb).

The registration of listeners depends on what type of reception technology the provider chooses to adopt. The listeners are still technically anonymous because there is no return signal from the receiver to the satellite. Registration can only come about by using a subscription system alongside an encrypted signal beamed from the satellite. By subscribing to Sirius, for example, you get a code that opens your receiver for the Sirius signal. Notice that the registration process is not internal to the platform as with broadband Internet and mobile telephony; it has to go via the post, telephone or e-mail.

Broadband Internet

The speech functionalities of broadband Internet are not very well developed compared to its graphical functionality. Indeed, the Internet was for decades a writerly medium and writing, photographs and graphical designs were the primary means of communication (Gauntlett 2000). The Internet is still largely organized so that users have to access a website or software in order to launch sound content. All types of sound are somehow secondary qualities of the Internet; at least they were so until the mid-2000s (Johnson, 1997). But now the Internet increasingly harbours live speech with or without video, some of which resemble broadcasting (YouTube) and others which resemble phone conversations (Skype). In the future, it seems that broadband Internet will provide a series of platforms that combine audiovisual communication with global reach and interactivity; and this combination has great utility as well as novel aesthetic qualities (see Leandros 2006).

The participation techniques involved can be categorized according to three perceptual foci, and these are central to all platforms described for broadband Internet (Johnson 1997). Firstly, there is participation with mouse and keyboard where users browse, select, store and organize all kinds of information that is displayed on the screen. Secondly, there is participation in writing with users producing all kinds of texts on e-mail, chat groups, personal HTML-programmed sites, public arenas like www.bbb.co.uk, and so on. And thirdly, there is participation in speech/music. The user can record sound and publish it on websites, or send voice messages and engage in live conversation in MSN Messenger,

Skype software and other audiovisual services. YouTube became a popular example of this in 2006–2007 and it is crucial to the participatory perspective of this chapter.

The terms of registration seem quite liberal. When sending messages over the Internet, many people enjoy a sense of anonymity by acting under pseudonyms and nicknames. Since there is no obvious way to associate these nicknames with a real world identity, users can create fictional identities and life stories for themselves on dating websites, political activist websites and newsgroups. But, despite this appearance of anonymity, it is easy for webmasters to identify the IP-addresses of computers, store information about their geographical location and trace the content of the hard drives. The provider of broadband access can easily shut off the connection to a specific IP-address, for example, if the user does not pay the bill on time. A determined party such a government prosecutor, a plaintiff in a lawsuit or a determined stalker may be able to identify almost any individual's PC, especially if they are assisted by the records of the Internet Service Provider that assigned the IP address in the first place.

4. Web radio

Internet radio, i.e. live streaming on the Internet, was first introduced in 1993 (Priestman 2002). Earlier users had to first download the sound file and then listen to it, but the stream player made it possible to listen to the sound while receiving it and, in this way, it recreates the 'liveness' of broadcasting on a completely different platform. The reception is not very mobile since it typically occurs on a stationary or portable computer. Wireless broadband reception is about to improve the mobility of Internet radio.

Web streaming does not introduce any improvement in the conditions for active public participation, but it lowers the threshold for establishing new editorial outlets, and opens new avenues for public participation in speech and music. This can include student radio stations (Coyle 2000) or more experimental communication (Nyre 2006; Delys and Foley 2006). Web radio services can be launched for small groups scattered all over the world without large initial investments. It has become the main delivery system for thousands of web-only radio operators and an important supplementary platform for practically all radio broadcasters. An editorial unit can target Internet services to a selected region using IP-addresses because it includes national co-ordinates. This 'directionality' gives Internet radio providers a great level of control – if they have reason to make use of it. BBC was contemplating shutting out foreigners from their domestic programme on the web and would have blocked all IP-addresses not originating in the UK.

Regarding the techniques of listening, users can do more than listen blindly to the sounds – they can also browse the website and operate the media player according to a host of functionalities that can be customized to their needs. The user can navigate the global menu of radio stations and music jukeboxes that are on offer. There is also the possibility to use the website actively for personal information needs, not just to read about programmes and stations, but to check the weather, the traffic situation and other practical information that is peripheral to the station's audio service. Regarding the terms of access set by editorial units and providers, web radio allows as much anonymity as the Internet can give. As already suggested, the location of the computer can be identified and activities on the computer can be monitored, as would be the case for any other activity on the Internet. Web radio is the Internet's equivalent to analogue radio.

5. Podcasting

Podcasting is distinctly different from web radio because it communicates exclusively in the form of pre-recorded programme packets. There is no live speech and no immediate actuality to the journalism that can be made, but there is a lot of speaking from celebrity hosts, news journalists and experts of many kinds (Berry 2006). Podcasting combines the mp3-format with RSS (Really Simple Syndication) feeds, and it is often listened to on music players. Earlier, a user had to search and download every audio file separately to his computer and transfer the files into the portable player, but from 2005 podcasting made it possible for users to subscribe to audio content. Whenever a website publishes new material through the RSS feed, the listener's software client automatically downloads it (Levy 2006). Typically, the user will transfer the files to his mobile mp3-player and listen while roaming around his geographical environment. This comes as a further contrast to web radio, which is more strictly bound to the more stationary setting of the computer.

Podcasting provides a series of private consumption techniques for citizens. It allows users to create a personalised series of subscription feeds from a range of providers, and since the programmes are downloaded automatically, the user automatically owns a hard copy of each programme. In the same way as web radio, the user can browse and read background information on the Internet while listening to the programme. It is worth noting that the podcast listener can easily become a podcaster. The software for publishing podcasts is just as easily available as streaming audio software, and this raises the opportunity for the media user to also be a content provider. This is a reflection of the Internet's great potential for symmetrical communication, but notice again that all these opportunities come at a cost. The technical (and editorial) providers can identify the user-terminal and all the data traffic on the terminal by the IP-number, and the people who publish podcasts through an ordinary broadband connection are therefore always monitored and potentially censored by their Internet service provider.

6. Music file-sharing

Like podcasting, music file-sharing deals with already recorded material and cannot communicate in real time. Among music lovers, players like iTunes, RealPlayer and Windows Media Player are the natural way of listening to music, with everyone finding and filing their favourites by the use of music software (Alderman 2001; Jones 2002). Music file cabinets give easy access to recorded music and it can be stored on the wearable iPod Nano just as well as on the stationary PC. The skilful user can find a certain track in five seconds maximum, and music collectors can stockpile as much as 20,000–30,000 songs in their file cabinet (Sterne 2006).

The piracy issue is only one dimension of the freedom of action that users have on a file-sharing platform, which is quite unlike web radio and podcasting. The music-lover can publish music files at will, without the prior agreement of the editors of the site or the musicians. The music will be downloaded and enjoyed by countless users who can write recommendations to each other to accompany the music files, for example, using the chat function in TorrentSpy. Users can create personalised play lists; they can access background information about artists and music, and they can purchase on-demand services that one actively selects in the Internet marketplace. There are so many providers of music and video files on the Internet that listeners are never dependent on one provider in order to get a certain type of music, which is again very different from Internet radio. If you want to listen to BBC Radio One there is only one provider.

The terms of registration set by editorial units and providers are more precise than those of web radio. First, there is the standard registration of the computer by the IP-number. In addition, it is not uncommon for file-sharing sites to require usernames and passwords, so that the content is restricted to a select group of people (Hacker 2000). Secure sites can retaliate against abuse or non-payment by instantly shutting off the music lover's access. File-sharing sites can log the user's individual preferences as they aggregate over time, and offer custom-made menus of music based on this information. Amazon.com has done this with their book sales for many years. It is difficult for the consumer to know that all the preferences they disclose will not be exploited for sinister purposes.

7. Social networking sites

In recent years, various audiovisual Internet services have added weight to the phenomenon of 'personalised media'. For example, music sites like www.pandora.com and www.last.fm allow users to tailor their automated musical output to their own genre tastes. Such personalisation is constructed by having huge numbers of songs available and offering ways of sifting through and cherry-picking favourites, all of which are stored in the provider's database.

But there are more potent examples of public participation on the Internet, and social networking sites like MySpace and Facebook are among the most prominent. Here, people can cultivate a personal profile with contacts, an archive of pictures, videos and texts, plus all kinds of ad-hoc or regular contact with other members of the site. The distinctive feature of these sites is that the feedback and responses from other users are important for all the content producers. For example, the popularity of videos on YouTube is measured not only in the number of hits and views, but also with a rating system that all users can score the videos by. There is a strong element of trust and credibility (and the lack thereof) between the users, and the editorial unit of traditional broadcasting is almost entirely bypassed. Twitter is the latest in a series of such social networking sites, and it is special in that it only allows 140 characters (like an SMS message), and can be hooked up to your mobile phone so that you can 'twitter' directly to the Internet site.

The differences between social networking sites and analogue broadcasting are so great that the two platforms are almost incomparable. The most notable differences are that the user can become his own editorial unit and publish audiovisual content at will, and that the user can write messages that are displayed on the screen, and organize and co-ordinate his network on his personal profile page. It almost goes without saying that the user can also create various types of personalised playlists based on the content accessed on such sites, and also read background information on the general Internet, and potentially buy commodities based on this multi-platform engagement.

As always, this great affordance of public communication comes at a cost. The terms of access set by editorial units and providers override all the interfaces for the users. This is brutally demonstrated by the fact that the webmaster can instantly remove any part of the content uploaded by the user, and also shut down his access to his personal profile or give the user a penalty quarantine period; since after all, the provider controls the usernames and passwords for the user profiles. It goes almost without saying that the provider can keep track of the user's entire media consumption history on their site, and can also identify the broadband subscriber through the ISP and track the location of the terminal by the IP-number of the computer.

Mobile telecom networks

The mobile phone has traditionally been a medium for personal contact more than journalistic content, and at first glance it may not seem to share many features with traditional broadcasting. People essentially use the mobile phone as an extension of their social lives. Anybody and everybody can speak on the phone. It is a private realm of communication where slang, codes, jive, lingo, accents, and dialects of all kinds are cultivated (Ling 2004). What made GSM more successful than other systems was that it was designed to bring mobile telephony to all people, instead of only business people (Manninen 2002: 298–302).

Mobile phone systems consist of a grid of transmission stations and reflectors, which create an overlapping pattern of access zones so that, in practice, the user can roam about freely and without caring about where he is. The provider registers the geographical information along with the time of any calls made. If necessary, the location of the phone equipment (and by extension the user) can be traced in detail, and the provider can also register all the numbers he or she has called. Notice however that users are well aware of these registration features, and can use pre-paid cards and dispose of phones regularly if they really want to stay anonymous (Gow 2005). Twitter must be mentioned again because it combines broadband Internet with SMS-messaging, and as such, allows providers two ways to track the activities of the users.

In contrast to broadcasting, the platform for telephony did not traditionally contain its own editorial institutions. But increasingly traditional journalistic content can be accessed through, for example, FM-equipment embedded in the mobile phone, and WAP and 3G services that present text information on the display screen. Notice that there are of course a host of services that channels mobile phone use into broadcasting, such as phone-ins on radio and various forms of SMS-messaging to television shows (Enli 2007; Hill 2005; Siapera 2004). It seems that the mobile phone still mainly functions as a feedback loop for the other platforms discussed in this chapter. Nevertheless, there is potential for editorial innovation when it comes to public services for mobile telecom platforms (Rheingold 2002).

8. 3G mobile audio and video

The Universal Mobile Telecommunications System (UMTS) can deliver audiovisual signals besides telephone calls and text messages, and opens up the telephone to all kinds of miniaturized radio and television services. The mobile phone, however, has a small screen and the audio must be heard through headphones, restricting its perceptual richness. Nevertheless, 3G phones make broadcast sound practicable within the mobile network itself, and not just via FM or DAB transmission (Farnsworth and Austin 2005; Lillie 2005). 3G mobile television is typically referred to as a 'unicast' service, as there is a dedicated signal stream to every single user. The problem with 3G-based Mobile TV services is that the network in a given location has a fixed capacity and only a limited number of individual connections are sustainable at any one time. This is not currently a problem but if the popularity of mobile TV were to increase, there would be a potential bottleneck here (Blandford 2007). Web radio also has limitations due to the principle of dedicated signal streams.

It is no surprise that web radio is present on UMTS. In March 2005, the UK-based radio broadcaster Virgin Radio launched its live radio services on the Internet for its 3G mobile phone users. Podcasting is also implemented on UMTS, and the Nokia Podcasting application allows for downloading podcasts with the N-series devices. Users can choose between Wireless LAN or a package data plan. An increasing number of phones have WLAN support as standard equipment (for example Nokia N91 or N95), which makes it possible to avoid the expensive data transfer via 3G network and not lose out on the functionalities.

When it comes to the balance of registration and participation, it is worth noting that 3G phone users can access all the Internet services analyzed above, including web radio, file-sharing, podcasting and social networking sites. In combination with the camera, microphone and video function which most phones have installed, this allows users to be mobile content providers and to publish their preferred stories about whatever social situation they choose, ranging from the work setting to deeply private settings. In addition, future 3G phones will in all likelihood be equipped with digital broadcasting receivers, particularly DVB-H. This adds to the FM and Internet-only services that are already available. At the end of this impressive list of bundled services, it must be mentioned that most mobile phones have good music players where users can create play lists and organize on-demand listening.

We have already discussed the registration features of the mobile phone, and here we will only add that the provider can shut off the user's access (due to non-payment or censorship); keep records of the individual's entire media consumption history on his device; and record all audio/video feeds to and from the user's device regardless of whether it is private or public. Again, the great range of techniques for participation cannot come about without an increasing range of control techniques among providers.

9. Location-based audio and video

What we call location-based services can, for example, be a mobile phone with facilities for satellite communication. Many mobile phones are equipped not just with GSM telephony and 3G broadband, but also have GPS satellite functionality, e.g. for in-car navigation. GPS is an American positioning system and there is also a European system called Galileo, and in the future there will be Russian and Chinese satellite communication systems. (GPS and Galileo are fully compatible so navigators can use data from both satellite systems). Each system requires several satellites that transmit their own positioning data. The device on the ground receives data simultaneously from several satellites, and thereby triangulates the position to the degree of metres and in three dimensions. Notice that the terms of access set by the providers are just as extensive as for the 3G phone (see above). The question is whether the new functionalities are worth the price of registration.

Regarding the techniques for public participation by citizens, the location-based platform is definitely the most versatile and powerful. The handheld device can contain all the platforms previously discussed (for example, FM, DAB and broadband Internet sites on 3G) and, furthermore, the location-information can be exploited for various purposes.

Until now, location-specific prototypes or experiments have mainly dealt with annotation of text messages to a certain place. Among the many interesting experiments in this field are the Geonotes project (Persson et al. 2002) and the Stick-E Note architecture (Pascoe 1997). There are also journalistic location services like www.everyblock.com, which presents news, current affairs and cultural events related to precise locations in American cities. However, there is still little or no sound communication available in these set-ups. At present, mainly museums and zoological gardens make use of such designs to present information about artworks or caged animals, typically on handheld devices that the spectator carries around with him/her in the building. Such localised information is not really public since only paying visitors have access to it.

One opportunity is to mount transmitters for wireless broadband (or Bluetooth) in public spaces and link this up with GPS positioning. The device will not only tell you your location, it can also disclose this location to the providers of the service and, by extension, may potentially be used for editorial purposes. While satellite radio sends the same signal to all locations, the GPS-driven service can have different signals to all locations, and this increases the potential for journalistic and participatory innovation.

Location-specific communication based on GPS-positioning and wireless broadband can be restricted to just a small area in the urban environment (one square kilometre, for example). Everybody who passes through the area can get access to the programme, and when they move beyond the outer perimeter, the programme is shut off. Providers can do this by combining the GPS-positioning system with a data transfer system. Alternatively, the cells and location information of the normal GSM network can be used to triangulate the position of the users, and this takes place to a level of accuracy of hundreds of metres.

Conclusion

The chapter has demonstrated the strong connection between increased participation among audiences and increased registration of their personal information by providers. All the platforms that have really extensive participation allowance also have really extensive control features for the providers. At the heart of this connection is something that sounds like a law: two-way communication in digital media requires personal registration of all communicators.

The way that most governments deal with this type of dilemma is by practicing 'technology neutrality' in regard to future broadcasting platforms. In the USA, in particular, the idea is that the market should decide for itself which technologies are best suited for broadcasting and other branches. But there is a potential problem about practicing technology neutrality. What if market selection does not favour public interest? The market winners may not possess the system with the best balance of registration and participation features. History is full of examples of good solutions that are ignored by the market. For example, in the early 1930s, the board of RCA wilfully ignored the FM-transmission system, delaying the introduction of a superb sound quality system and more channels in the USA for several decades (Douglas 1999: 256–267). Unless governments impose regulations towards specific solutions, the strongest business interest will prevail in the end. In the European context, it is therefore interesting that in 2007 the European Commissioner for Information Society and Media stated that the industry should agree on one single standard, and that it should be the DVB-H family of standards. Viviane Reding also suggested that if the industry and member states failed to agree on one standard, she would be forced to intervene with regulatory measures (Reding 2007).

Our comparison is not technology-neutral, but our conclusion is nevertheless ambivalent. We do not think that DVB-H is the better platform since it has extensive registration without extensive public participation. None of the next generation platforms lacks registration with rich participation opportunities; they are good at either one or the other. The better platform, regarding registration, would be a terrestrial universal access radio system like DAB, since it secures absolute anonymity for users. But the downside is, of course, that it does not provide any new interfaces for participation. The better platform regarding participation would be the location-aware type (on 3G phone networks), but it gives the provider such an extensive overview of the users' personal activities that it threatens their right to privacy. It seems that the citizen cannot have it both ways: they have to choose either unregistered access or wide-ranging participation opportunities.

References

Ala-Fossi, M. (2005a), *Saleable Compromises – Quality Cultures in Finnish and US Commercial Radio*, PhD Dissertation, Tampere: Tampere University Press.

Ala-Fossi, M. (2005b), 'Mapping the Technological Landscape of Radio: Where do we go next?', Paper at the *First European Communication Conference*, Amsterdam, 24–26 November 2005.

Ala-Fossi, M., S. Lax, B. O'Neill, P.Jauert and H. Shaw (2007), 'The Future of Radio is Still Digital – But Which One? Expert perspectives and future scenarios for the radio media in 2015', *Journal of Radio and Audio Media*, 1, 2008.

Ala-Fossi, M. and A. Stavitsky (2003), 'Understanding IBOC: Digital Technology for Analog Economics', *Journal of Radio Studies*, 10:1, pp. 63–79.

Alderman, J. (2001), *Sonic Boom. Napster, mp3, and the New Pioneers of Music*, New York: Basic Books.

Beniger, J. (1991), 'The Control Revolution', in David Crowley and Paul Heyer (eds.), *Communication in History. Technology, Culture, Society*, New York: Longman.

Berry, R. (2006), 'Will the iPod Kill the Radio Star? Profiling Podcasting as Radio', *Convergence*, 12:2, pp. 143–162.

Blandford, R. (2007), 'Nokia's second DVB-H Mobile TV handset – the Nokia N77', http://www.allaboutsymbian.com/features/item/Nokia_N77_Preview-DVB-H_Mobile_TV_Handset.php. Accessed 19 March 2007.

Briggs, A. and P. Burke (2005), *A Social History of the Media: from Gutenberg to the Internet*, Cambridge: Polity Press.

Bull, M. (2000), *Sounding Out the City. Personal Stereos and the Management of Everyday Life*, Oxford: Berg.

Crisell, A. (1994), *Understanding Radio*, London: Routledge.

Coyle, R. (2000), 'Observations from an Experiment in "Internet radio" ', *Convergence*, 6:3, pp. 57–75.

Dahl, H.F. (1999), *Dette er London. NRK i krig 1940 – 1945/ This is London. NRK at war 1940 – 1945*, Oslo: Cappelen.

Delys, S. and M. Foley (2006), 'The Exchange: a radio-web project for creative practitioners and researchers', *Convergence*, 12: 2, pp. 129–135.

Douglas, S. (1999), *Listening In. Radio and the American Imagination*, New York: Times Books.

Enli, G. (2007), *The Participatory Turn in Broadcast Television. Institutional, Editorial and Textual Challenges and Strategies*, Oslo: Unipub.

Farnsworth, J. and T. Austin (2005), 'Assembling portable talk and mobile worlds: sound technologies and mobile social networks', *Convergence*, 11: 2, pp. 14–22.

Fernback, J. and Z. Papacharissi (2007), 'Online privacy as legal safeguard: the relationship among consumer, online portal, and privacy policies', *New Media and Society*, 9: 5, pp. 715–734.

Gauntlett, D. (ed.) (2000), *Web.Studies: Rewiring Media Studies for the Digital Age*, London: Arnold.

Gow, G.A. (2005), 'Information Privacy and Mobile Phones', *Convergence*, 11: 2, pp. 76–87.

Hacker, S. (2000), *MP3: The Definitive Guide*, Sebastopol, CA: O'Reilly.

Hendy, D. (2000), *Radio in the Global Age*, Cambridge, UK: Polity Press.

Hill, A. (2005), *Reality TV. Audiences and Popular Factual Television*, London: Routledge.

Jauert, P. and G.F. Lowe (2005), 'Public Service Broadcasting for Social and Cultural Citizenship. Renewing the Enlightenment Mission', in Lowe and Jauert (eds.), *Cultural Dilemmas in Public Broadcasting Service*, Gothenburg: Nordicom.

Jenkins, Henry (2008), *Convergence culture. Where old and new media collide*, New York: New York University Press.

Johnson, S. (1997), *Interface Culture. How New Technology Transforms the Way We Create and Communicate*, San Francisco, CA: HarperEdge.

Jones, S. (2002), 'Music that moves. Popular music, distribution and network technologies', *Cultural Studies*, 16: 2, pp. 213–232.

Katz, J.E. and M. Aakhus (eds.) (2002), *Perpetual Contact. Mobile Communication, Private Talk, Public Performance*, Cambridge, UK: Cambridge University Press.

Lax, S. H. Shaw, M. Ala-Fossi and P. Jauert (2008), 'DAB – the future of radio? The development of digital radio in four European countries', *Media, Culture and Society*, 30:2.

Leandros, N. (ed.) (2006), *The Impact of the Internet on the Mass Media in Europe*, Bury St. Edmunds: Arima Publishing.

Lessig, L. (1999), *Code and Other Laws of Cyberspace*, New York: Basic Books.

Levy, S. (2006), *The Perfect Thing. How the iPod Shuffles Commerce, Culture and Coolness*, New York: Simon and Schuster.

Lillie, J. (2005), 'Cultural access, participation, and citizenship in the emerging consumer-network society', *Convergence*, 11: 2, pp. 41–48.

Ling, R. (2004), *The Mobile Connection. The Cell Phone's Impact on Society*, Amsterdam: Elsevier.

Manninen, A.T. (2002), *Elaboration of NMT and GSM standards. From Idea to Market*. Doctoral dissertation, Studia Historica Jyväskyläensia 60, Jyväskylä: University of Jyväskylä.

Meyrowitz, J. (1994), 'Medium Theory', in David Crowley and David Mitchell (eds.), *Communication Theory Today*, Cambridge (UK): Polity Press.

Murdock, G. (2005), 'Building the Digital Commons. Public Broadcasting in the Age of the Internet', in G.F Lowe and P. Jauert (eds.), *Cultural Dilemmas in Public Service Broadcasting. RIPE@2005*, Stockholm: Nordicom.

Nyre, L. (2007), 'Minimum Journalism. Experimental procedures for democratic participation in sound media', *Journalism Studies*, 8: 3, pp. 397–413.

Nyre, L. (2008), *Sound Media. From Live Journalism to Music Recording*, London: Routledge.

Nyre, L. and M. Ala-Fossi (2008), 'The Next Generation Platform: Comparing Audience Registration and Participation in Digital Sound Media', *Journal of Radio & Audio Media*, 15:1, pp. 41 – 58.

Pascoe, J. (1997), 'The Stick-e Note Architecture: Extending the Interface Beyond the User', in J. Moore, E. Edmonds and A. Puerta (eds.), 1997 *International Conference on Intelligent User Interfaces*, ACM Digital Library, ACM, pp. 261–264.

Persson, P. et al., (2002), 'GeoNotes: A Location-based Information System for Public Spaces', in K. Höök, D. Benyon and A. Munro (eds.), *Designing Information Spaces: The Social Navigation Approach*, Springer, pp. 151–173.

Priestman, C. (2002), *Webradio. Radio Production for Internet Streaming*, Oxford: Focal Press.

Reding, V. (2007), 'Mobile TV: The time to act is now', http://europa.eu/rapid/pressReleasesAction.do?reference=SPEECH/07/154&format=HTML&aged=0&language=EN&guiLanguage=en. Accessed 25 May 2009.

Rheingold, H. (2002), *Smart Mobs. The Next Social Revolution*, Cambridge, MA: Basic Books.

Rodman, G. and C. Vanderdonckt (2006), 'Music for nothing or, I want my mp3: The regulation and recirculation of affect', *Cultural Studies*, 20: 2–3, pp. 245–261.

Siapera, E. (2004), 'From couch potatoes to cybernauts? The expanding notion of the audience on TV channels' websites', *New Media & Society*, 6:2, pp. 155–172.

Sterne, J. (2006), 'The mp3 as cultural artifact', *New Media & Society*, 8:5, pp. 825–842.

Tolson, A. (2006), *Media Talk. Spoken Discourse on TV and Radio*, Edinburgh: Edinburgh University Press.

Winston, B. (2005), *Messages. Free Expression, Media and the West from Gutenberg to Google*, London: Routledge.

Ytreberg, E. (2004), 'Formatting participation within broadcast media production', *Media, Culture & Society*, 26: 5, pp. 677–692.

Chapter 11

The Online Transformation: How the Internet is Challenging and Changing Radio

Helen Shaw

The promise of digital technology to both producers and users was to make things better, to enhance the experience of radio. In September 1995, Liz Forgan, then BBC's Director of Radio, hailed it as 'the third age of radio' (Williams 1995). The rhetoric of terrestrial digital radio in the mid-1990s, whether DAB or any other platform, was to add value: extra services, better sound, improved choice and quality.[1] When listeners were asked what they wanted, which was not very often, they frequently said they wanted to be able to find their favourite programmes, listen to them again or hear a station which was only on Short Wave or Long Wave. Many mentioned extra choice, different stations, niche music or being able to access their local station, particularly if they were living abroad or in a different city.

Digital radio technologies have met some of those promises, but part of the conflict in the match between promise and delivery is the limitation of terrestrial broadcasting, particularly DAB, to its own spectrum footprint, and equally its ability to match the user's often very basic desire to hear their favourite radio when and where they want it, to find things easily, save it and listen to it again. Up until the advent of podcasting, users had to request and often pay for a cassette tape (or later a CD) copy of a programme if they missed it, and many services did not record or archive all their output. Radio was essentially ephemeral.

In a sense, what has been happening since 1995 is a twofold digital radio transformation: one top down, led by terrestrial broadcasting platforms like DAB, DRM, DVB-H etc; and the other a bottom up surge led by the emergence of broadband-enabled Internet. The top down approach has been costly, limited by the need to get a network in place, new appliances in the users' hands and a fragmented approach to platforms. The bottom up one has been defined and led by users seeking out content and beginning to make their own. The end result is that by the end of 2010, many traditional radio stations in Europe may have stronger audiences online through live streaming, catch-up services, podcasting and downloads, than through terrestrial broadcast digital platforms. Ultimately, the dominant platform will still be FM long into the future.

The outcome is less about a failure in the Eureka 147 ideal of a common European platform, than in a failure to match digital solutions to the often simple questions and needs being raised by the audience. The marriage of radio and the Internet has, in the short term, met more of those needs, and allowed radio to flow into the personalised media devices such as laptops and mobile handsets. It has globalised radio through Internet radio sets and reduced linear and geographical barriers. It has not, and can not, replace the full broadcast network; radio still needs a digital terrestrial network which will future proof its market. The benefits of the old European ideal of a common shared platform are real not just for audiences, but for the stability of a market economy supporting radio content. But if a digital terrestrial platform is to challenge the online lead it may need to perceive technology through the eyes and ears of the users – and imagine what it can offer them. The end box, the radio receiver, is going to have to combine the best of both analogue and online radio, mixing linear and non-linear. Just like the promise of the 1990s, digital radio has to be better, has to improve the FM package, or it is not going to compete with it, let alone overtake it.

The Birth of Pod Radio

On May 16th 2005 a failed AM talk show radio channel in San Francisco made headlines when its owners, Infinity Broadcasting, converted it into a listener-submitted podcast-only radio station and called it KYOURadio. Just over a year before, on February 12th 2004, a writer in *The Guardian* newspaper, Ben Hammersley, suggested the word 'podcasting' to describe a type of audio-blog like those of Adam Curry and Christopher Lydon, which could be downloaded using an RSS (Really Simple Syndication) feed and played from a PC or exported to an MP3 player. Across 2004–05, the new wave of podcasting was seen as an attack on the old AM/FM radio model with its linear and scheduled flow of talk and music, and media journals were full of stories headlined 'podcasting kills the radio star' with or without a question mark.[2]

Curry, the self-acclaimed godfather of podcasting, saw it as the triumph of amateur content over big media: 'it's totally going to kill the business model of radio' he said in October 2005. In his view, podcasting would kill radio's commercial advertisement business model by effectively creating a network of audio niches where users could skip the adverts. Some mainstream radio broadcasters held back from podcasting on the working assumption that it would erode their traditional mass media base, fragment their primetime audiences and lose them spot advertisement revenue. By mid-2009, the fear of cannibalising its own audiences was still stopping some European local radio stations from offering podcasts.[3] The arrival of KYOURadio seemed like the industry's worst nightmare: a station broadcasting listener-generated content uploaded through the station's website.

Yet what was happening was two-fold. User-generated, amateur audio shows were growing in number and popularity, but in tandem with that bottom-up trend was the push down by global radio brands, like the BBC in the United Kingdom and NPR in the United States, who began moving into podcasting and re-distributing their own shows or edited highlights of their shows, in podcast pilots from late 2004. By 2005, the BBC had launched full-scale podcast pilots across its network from its youth orientated music channel, BBC Radio 1, its sports channel Five Live, to its more middle-aged speech channel, BBC Radio 4. By the end of 2005, it was already seeing significant take-up not just from the young, selecting music and comedy podcasts from BBC Radio 1, but also from its 45+ audience, who soon made the BBC Radio 4 daily breakfast news show, *Today*, and one of the station's erudite presenter-led shows, *In Our Time* with Melvyn Bragg, amongst the most popular BBC podcasts on iTunes (Marriner 2006; Shaw 2005).

What drove the rapid take-up, and what listeners loved, was that here was a technology which actually improved access to radio. Podcasting allowed radio to time-shift and free itself from the rigidity of the station schedule. It gave listeners not just a 'listen again' facility, which they already had on the Internet, but the ability to control when and where they heard, shared and stored their favourite shows. It equally gave them the ability to be their own radio programmer, mixing podcasts from different genres, stations, even different countries, and building their own schedule. TV had had the VCR long before digital programming, but radio had always been viewed as ephemeral – 'writing on the wind' as one Irish broadcaster called it (McRedmond 1976) – with no easy tools for pre-recording, saving or finding old shows.

For radio fans, while the move to podcasting led to a shift away from linear to non-linear radio, it was never about rejecting 'live' radio. It was simply about opting for the choice of having radio on-demand whenever you wanted it rather than tied to a programmer's schedule. If Google answered the global need to find things, to put order on information, then the same process is now underway with audio and video, allowing you to find and hear what you want, where and when you want to (Marks 2006).

While Ben Hammersley helped give life to the term 'podcasting', it had little intrinsically to do with either Apple's iPods or broadcasting, and it was not until July 2005 that iTunes began offering podcasts as an option. Up until then, podcasts were distributed through portals like Curry's ipodder.org and from individual websites. Once iTunes put podcasts in its global window podcasting took off at a phenomenal rate, not just in terms of downloads but in the number of both amateur and professional shows being released. iTunes is still the dominant delivery platform for podcasts, with as much as two thirds of all podcast feeds coming from it (RAJAR/MIDAS 2009; Arbitron/Edison 2008).

By the autumn of 2005, more and more international and national radio stations were exploring podcasting and beginning to see it not as a threat, but as an opportunity. National Public Radio (NPR) in the United States began a limited podcasting pilot on August 31st and by November had over 5 million downloads, rapidly altering its production cycle and thinking to expand to meet the online media demand (Shaw 2005; Glaser 2005). By the beginning of 2006, it was becoming clear that far from killing off radio, podcasting was offering traditional radio another route into digital distribution and one that it could use to its advantage to survive in the digital age.

One issue was that this new online route, broadband-enabled audio (via downloads or RSS feeds) would involve traditional radio broadcasters having to re-think their definition of radio, and adapt their approach to fit the new environment. In the end, the twin tracks of podcasting – the listener/user generated content which saved KYOURadio, and the big brand shows of public and commercial broadcasters – could both thrive and find ears. The challenge remains the one that Curry posed in 2005 in terms of radio economics, of trying to find a new business model to suit convergence radio (Green, et al. 2005).

Podcasting and listener research[4]

Qualitative audience research on podcasting has been rare, but since Richard Berry in 2006 posed the question – 'Will the iPod kill the Radio Star?' (Berry 2006) – we can refer to a couple of examples that give an insight into some of the special qualities users have found in podcasting, both as an activity in itself and related to the use of other forms of radio.

Two Danish qualitative audience research projects from 2006 have examined user patterns and preferences among radio listeners to DR (Danish Broadcasting Company), which launched podcasting in 2005 with an immediate and fast growing interest from a wide range of listeners, especially heavy users of P1, the culture and current affairs channel (Ala-Fossi and Jauert 2006). In the same year, Berry raised the question whether podcasting should be considered a specific medium or a new form of radio distribution (Berry 2006). Berry would leave it to users to decide, but from the two Danish studies both aspects are included: it is radio (since most of the podcasts are offered by radio broadcasters), but its specificity as a mobile medium adds a new dimension to the act of listening, a more individual or personal dimension.

In the first research project from 2006, fifteen qualitative interviews among podcast and radio listeners from 21–75 years old were conducted. These focused on three main aspects: adoption or what incentivised the listener to start using podcasts, content preferences, and how podcasts were used in relation to time, space and user context (Thomsen and Starklint 2006). Among the participants in this study podcasting adoption was prompted by existing radio content preferences, combined with a curiosity towards (new) technology. The main provider was – and still is – DR, but younger people, in particular, had a greater interest in getting podcasts from other sources (Thomsen and Starklint 2006: 115).

The study revealed three types of user profiles defined from their content preferences: the technology oriented user, who acts a 'first mover', trying out new devices and new opportunities in the market; the independent programme planner has a more pragmatic approach to podcasting, considering it a more accessible shortcut to preferred radio content to be used 'on demand'; and the selective user who combines the easier access to already familiar radio programme content (DR), with an interest in added value content, mainly to be found as links on the DR website to podcasts from other sources, or to additional radio archive material etc.

Podcasting is first and foremost a mobile medium to be used any time and anywhere. The mobility offers independence from the traditional time schedule of radio, and also a chance to use your personal dead time when you are stuck commuting, cleaning, cooking, or just walking the dog. But most of the respondents emphasised podcasting as the creator of 'a personal soundtrack to your everyday life', a phenomenon also known from Bull's analysis of mobile use of music and MP3 players, where you can escape and create your own playlist (Bull 2000).

From the results of this Danish qualitative study, radio as we know it does not seem to be in much danger. You may argue that through podcasting radio has opportunities to strengthen its reach to even more segmented groups, depending on the scale and character of the podcast supply. In Europe many PSB companies are ahead of their commercial competitors, mainly because of the music copyright issues involved in distributing music radio as downloads, combined with the lack of an effective business model for podcasting.

The second Danish audience research project was also conducted in 2006 by DR and Aarhus University among a group of 15–19 year old high school students. It was a combination of a quantitative survey with 355 participants, used as a recruitment basis for individual in depth qualitative interviews (12) and focus groups (2) (Kjær and Plom 2006). The aim of the project was to obtain knowledge of young 'early adopters' in podcasting about their motivation for using this new mobile medium – as a stand-alone medium and/or a distribution channel for DR radio programmes and audio products. For DR, the research project had a strategic objective, namely to establish a platform for the development of podcast content aimed at younger audiences since the initial breakthrough of DR podcasting in 2005 was used mainly by older audiences for arts, culture and fiction programming on P1, the culture channel. How to create non music content that could appeal to younger people was the strategic challenge facing the broadcaster.

Podcasting is an individual medium in its normal usage – a personal space simulating sociability and safety. For the young users in this study, podcasting was considered a 'family member' of other digital and social media: video streaming, web radio, Tivo etc. iPod and iTunes were also considered as interrelated and, for some, podcasts were thought to be proprietary Apple Macintosh products. Their preferred content is both exclusive and non-exclusive products. The exclusive, unique audio products include sports news, interviews and new mixed music genres (hiphop mixtapes or indiefeeds), and the non-exclusive are often known from other media portals like newspapers/magazines (Kjær and Plon 2006: 34–36). Favourite radio programmes for podcasting consist of talk shows and comedy programmes, and the combination of entertainment and information are the most attractive for this age group.

Hybrid Radio

'*The hybrid or the meeting of two media is a moment of truth and revelation from which a new form is born*' – Marshall McLuhan, *Understanding Media*, 1964

'But is it radio?' was often the response of European radio people in the midst of a discussion on digital radio, particularly when images were wedded with audio or the concept of radio coming to you other than through a broadcast network was raised. For many, particularly in European public broadcasting, radio means literally being broadcast on radio waves as a channel of continuous content. Yet radio has adapted to each technological change since its invention, from crystal sets and vacuum tubes to being liberated as a mobile medium with the marriage of transistors and FM in the 1970s and Sony's Walkmans in the 1980s. Its ability to time shift through the Internet has brought a new radio renaissance through streaming and the growth of Internet radio players, allowing people to shift locations beyond radio spectrum and literally hear words and music from across the globe, to 'listen again' through downloads and podcasting feeds. In a sense, the marriage of one of our oldest mediums, radio, and our newest, the Internet, has brought forth a classic McLuhan hybrid.

What the Internet has offered to radio, through live streaming and podcasting, is a means of breaking down the barriers to access in time, space, geography and form. It allows us to find a six week old programme from a station on the other side of the world and listen to it during a walk in the park or driving to work. We can find just what we want and ignore the rest – including the adverts.

For NPR, the US public broadcaster, the match has been ideal given its vast geographical base and its reliance on sponsorship rather than spot advertisement (Glaser 2005). The car manufacturer, Acura, came on board as sponsor of NPR podcasts from its initial pilot in September 2005. For sponsors, unlike advertisers, the returns are now greater and potentially more valuable given that their sponsorship is now reaching beyond its radio spectrum footprint, and also given that the download figures provide quite an accurate picture of who is listening. Ironically commercial broadcasters in the UK and Ireland who are dependent on an often declining spot advertisement pie, are looking increasingly to embedded commercial sponsorship for both live and podcast shows in order to re-fit their business model (Hirst 2009).

NPR now has up to twelve million podcast downloads a month and creates shows solely for online and podcast use, allowing it to expand its ability to serve parts of its diverse audience and provide added-value content, which enrich and grow its audience base. The website traffic is up 78% across 2008 and attracts over 8 million unique visitors a month. Yet, like other broadcasters, most of its podcasts are still being downloaded via iTunes, which continues to act as a podcasting gate-keeper. NPR talks of its listeners as a community, NPRistas, and is using social networking tools and principles to nurture that community, including adding a podcast player to its NPR MySpace site (Dorroh 2008). Linking content to interactivity tools, whether blogs, web-chats or programming allows the listeners to add their own stories, and thus lets NPR benefit from the 'me media' culture where people want to use digital tools and the Internet to create personalised communications, yet still ensuring that content flows back into its professional brand to strengthen it.

The New York Times CEO, Arthur Sulzberger Jr., has said he is not too worried if the print edition of the paper disappears; 'within our lifetimes the distribution of news and information is going to shift to broadband', he argued (Businessweek 2005; Shaw 2007). The New York Times is about delivering news not paper, and in the same way NPR, in its shift online, is defining its business as story-telling, predominately audio story-telling. But it will also use whatever means and methods to tell stories, including print and video. NPR is effectively transforming itself into a multimedia company with radio at the top of the tree, and its recent appointment as CEO of Vivian Schiller, the former general manager of

The New York Times online, is a core part of that process (Miller 2009). For example www. npr.org has a vlogging (video blogging) grassroots community and 'This American Life', one of Chicago Public Radio's most popular and eclectic radio shows, now has a global audience not just as linear and podcast audio, but also in short form video on YouTube, on broadcast TV and even as a 'live' theatre show!

While most radio podcasts are free, NPR does charge $0.95 cents for archive shows, and Chicago Public Radio, for example, offers 'This American Life' free as a podcast for a short period after live transmission, and then charges $0.95 cents or you can buy a CD. While the general trend remains that podcasts are provided free by broadcasters, the NPR archive model may become more standard, with micro-payments for archive podcasts or podcast-only series. John Hirst, Head of Global Content and Podcasts at Global Radio, one of the main UK commercial radio players, urged broadcasters to explore revenues streams like micro-payments with hit podcast shows, quoting the example of the London station LBC which had a small, paid subscriber base. But revenues from radio podcasts remain extremely low, and even Hirst himself claims his group had only generated an extra £1.5million podcast revenue over 3 years (Hirst 2009).

Podcasting radio: challenging the broadcasters

For NPR, going online helped their business model. For many European public broadcasters, the concept of moving audiences away from terrestrial broadcasting has been anathema. The DAB terrestrial digital radio path had matched traditional broadcasting ideology of broadcasters operating channels of continuous content to mass audiences, while online content downloads and podcasting meant filleting your schedule and parcelling out the bits to individuals. The grassroots success of non-radio audio bloggers like Adam Curry in the United States, and initiatives like the Ricky Gervais top rated comedy podcasts from *The Guardian* in 2006, had encouraged broadcasters to 'pilot' podcasting, but it was the audience's take-up which has forced broadcasting companies to re-think their digital audio strategies and begin seeing podcasting not as an add on tool to their websites, but as a core means of content distribution.

A challenge for broadcasters from the beginning has been the content rights issues related to offering speech and music programming as downloads rather than 'listen again', particularly in relation to programmes made by independent production companies, to sports content and to original music, drama and arts programming, which the copyright or intellectual properties of performers, composers and artists may be limited to a handful of linear broadcasts, with a prohibition on handing the content over to the audience. For broadcasters, this has often made using their archives problematic unless all the rights, including in all electronic forms, were signed over to the broadcaster. While many radio and TV channels now attempt to gain full intellectual property rights in commissions or acquisitions, it remains a significant and complex block to using all linear content in either a download or podcast format. The EBU Youth Radio survey (2008) showed that 82% of European public broadcasters were offering podcasts but only a third included music.[5]

For many broadcasters, the solution to this may be that rights based content, which has a limited on-air shelf life, will become paid content, with micro-payments returning value back to both the broadcaster and the artists/rights holder in the same way as book or CD sales. The challenge to this will remain the same as for music or film in the Internet age, where paid content is declining given the ease of sharing content for free online.

European public radio broadcasters and podcasting

As part of this chapter's research, we asked public broadcasters about their podcasting production to find out how they were adapting, what policies they were creating, what they knew about their podcast users and how they were funding the additional activity. We also tracked any research being conducted by the EBU Radio Department in Geneva. We focussed on European public broadcasters in order to compare similar organisations in terms of ethos, remit and funding. Equally, public broadcasters through the EBU have been at the fore-front of terrestrial digital radio developments, and in many countries have been tasked to be digital pioneers as part of their remit (European Broadcasting Union 2007). In some countries, commercial broadcasters are far more advanced in their use of online and interactive radio, while in other markets innovative leaders include non-broadcast players like newspapers, NGOs and new media groups.

Most European public broadcasters we surveyed began experimenting with podcasting by 2006, with the initiative usually anchored in the audio/radio division.[6] By late 2008, few countries had independently audited podcast audience research, except for RAJAR in the UK, and most broadcasters were re-versioning existing radio shows with some limited exploration of podcast-only production. In the United Kingdom, the BBC, an early adopter of podcasting, has been barred by its governing authority, the BBC Trust, from creating original, non-broadcast content for the Internet, as this is seen as outside its public broadcasting licence fee remit.[7]

It does however create 'best of' compilations of top news and entertainment shows, on a daily and weekly basis, and these podcast fillets are among the most popular BBC podcasts. For those without such a bar, the ability to use podcasting to pilot new ideas (as in NPR) or release shows which cannot find room in the schedule (RTÉ Radio 1's exam podcasts) has encouraged innovation. SWR in Germany creates 'podcast first' content, and DR in Denmark uses podcast shows as a means of connecting with younger audiences or providing added value content to broadcast shows. In the Netherlands, the public broadcasters have created www.radiocast.nl as a central portal for public radio podcasts, news and download data. Radio Netherlands Worldwide (RNW), the international service, releases podcasts of its schedule or edited segments, but it has also experimented with a podcast only show 'Media Network', which was a long running live radio show but is now a weblog of media developments.[8] In Finland, where the public broadcaster, YLE, began podcasting by September 2005, they have one audio show called 'Pop Talk' which originates as a podcast and is then broadcast on the radio schedule. They also have a video podcast, 'YLE Extras', with out-takes and funny on-air bloopers, which is solely distributed online.[9]

The BBC's first podcast pilot in November 2004 consisted of content from its Five Live station of news and sports. The full pilot from May the following year involved twenty programmes across a diverse range of content, and by July 2007 a full service was established, with over 200 podcast series by mid-2008 and some 300 by February 2009. The current service is audio only, despite a video podcast trial in 2006/7, and video's online offering is centred on the BBC's catch-up tool, iPlayer. The BBC is developing what it calls Radio Plus, which will provide its audience/users with a full, non-linear radio tool, allowing audiences to programme their listening, record programmes, select and gather programmes from across the services, and personalise one's own schedule.

In podcasting, what works for most broadcasters is short form content, particularly comedy podcasts, as well as edited highlights of top news/talk/current affairs shows, that allow people to catch up. Comedy, music and news/current affairs remain the top shows in most countries, across both public and commercial broadcasters. In Ireland, RTÉ creates some limited podcast only material, both add-ons, like full interviews, or exclusive content like its 'Radio Extra' show, and about 5% of its podcasts are video.[10] In its audience survey, it found a marked bias to male users (70% were men), but nearly 40% of its users were from outside the Republic of Ireland and, contrary to the US and UK research, it found as much as 51% of its users downloaded content directly from its website www.RTÉ.ie rather than iTunes.

The BBC releases its radio podcast download numbers on a monthly basis and has nearly 20 million downloads a month (Jan 09), with as many as 2 million downloads for its most popular shows like the BBC World Service Global News daily podcast and the BBC World Service documentary archive (see Table 1). BBC Radio 1's 'Best of Chris Moyles', a comedy and entertainment chat show, is regularly one of the most popular radio podcasts in the United Kingdom and gets over 2 million downloads for its two weekly podcasts. About half a dozen BBC radio podcasts receive between 500,000 to a million downloads a month, predominately comedy/entertainment, news/current affairs and arts, including BBC Radio 4's 'Best of Today' and its long running soap 'The Archers'.

With over 300 podcasts now on offer from the BBC, its total download number has grown significantly across 2008. In some cases, popular shows have doubled their take-up between August 2008 and February 2009, and while podcasting is, overall, dominated by the young, BBC Radio 4's consistently strong podcast performance shows that they are now a habit with ABC1 over-45 year olds. By July 2008, in his address to independent radio producers, Radio 4's Controller, Mark Damazer said podcasts were now an integral part of the station's relationship with its audience, and credits podcasting as one of the factors why Radio 4 was holding and increasing its audience.

Table 1: BBC Top of the podcasts Top 10 monthly downloads, Jan 2009.

1	BBC World Service Global News	1,969,061	(daily)
2	BBC World Service Archive Documentary	1,830,352	(several times a week)
3	BBC Radio 1 'The Best of Christ Moyles'	1,464,210	(weekly)
4	BBC News 'Newspod'	1,002,233	(daily)
5	BBC Radio 4 'Friday Night Comedy'	832,896	(weekly)
6	BBC Radio 1 'Best of Chris Moyles – Enhanced'	768,210	(weekly)
7	BBC Radio 4 'The Archers'	743,222	(daily)
8	BBC Radio 4 'Best of Today'	729,957	(daily)
9	BBC News 'Peter Days World of Business'	684,679	(several times a week)
10	BBC Radio 4 'In Our Time with Melvyn Bragg'	591,396	(weekly)

Source: BBC

By January 2009, BBC Radio 4 had increased its linear radio weekly reach to 12.4% (RAJAR Q4 08), reaching nearly 10 million listeners a week. While the global credit crunch and recession is seen as one of the primary reasons for the jump in the station's listening, it is clear that the substantial growth in non-linear listening, both through the BBC iPlayer and podcasts, has not hurt the station, but has helped deepen the reach of its often expensive speech programming.

Podcasting radio: changing the audience

In Ireland, the national public broadcaster RTÉ was a slow adopter of podcasting. While its big next door neighbour the BBC was running pilots from late 2004, RTÉ did not fully move into podcasting until July 2006. By the close of 2007, RTÉ claimed over 5 million podcast downloads, predominately of re-packaged radio shows. By March 2009, RTÉ was claiming 600–900,000 downloads a month, compared to an average daily linear audience of 1.2 million listeners on FM.[11] Like the BBC, its popular shows are comedy: 'Nob Nation' (RTÉ 2fm) and the national, breakfast news show, 'Morning Ireland' (RTÉ Radio 1) both pulling from quite different demographics; as well as high profile interviews from weekend shows like Marian Finucane, which often became 'must-hear' conversation pieces, like Finucane's interview with Irish author and her own best friend, Nuala O'Faoláin, who talked openly and honestly about her impending death from cancer in early 2008. Within days, over 60,000 people had downloaded the interview and it effectively put the Finucane show in the top ten iTunes charts – despite the fact the radio show largely appeals to a 45+ age group. The O'Faoláin, interview remains one of the most downloaded RTÉ interviews and still got regular traffic long after the popular writer had died. This rapid take up and success of podcasting stands in contrast to just thousands of listeners on the Irish DAB pilot which involved creating new, original material and channels like RTÉ Junior, Choice and Gold.

So what are people listening to? On March 28th 2009 RTÉ's website gave its top five radio podcasts as:

Table 2: RTÉ Irish Radio Top Five podcasts.

1	Nob Nation	2fm comedy
2	Marian Finucane	RTÉ Radio 1; an archive of current affairs interviews across the last year as well as the new editions
3	Playback	RTÉ Radio 1 'best of' compilation radio show
4	Colm and Jim Jim	2fm entertainment breakfast show
5	Eamonn Dunphy	RTÉ Radio 1's weekend interview based chat show

On iTunes the top Irish downloads, on the same date, were:

Table 3: iTunes Top Irish Podcasts.

1	The Emergency	Comedy show from Newstalk 106 national commercial station
2	Ian Dempsey	Breakfast comedy from Today FM, national commercial station
3	Nob Nation	RTÉ 2fm's breakfast comedy slot
4	Ricky Gervais	UK comedy podcast – non radio originated
5	Stephen Fry	UK comedy podcast – non radio originated

While independent audience research for digital radio and listening via the Internet is not yet available in small radio markets like Ireland, in the United Kingdom RAJAR (Radio Joint Audience Research) has been tracking audio listening via the Internet since the final quarter of 2007. These regular MIDAS (Measuring Internet Delivered Audio Services) reports, along with research commissioned by the BBC and separately by the communications regulator, Ofcom, give a good comparative view on how radio is being changed by podcasting and just who is listening and why.

By December 2008, the third RAJAR/MIDAS survey allowed us to see a one year trend in podcast usage in the United Kingdom. Across a full year, listening to radio via the Internet had grown to a third of the UK population (16.1 million compared to 12 million). But the more dramatic increase had been in podcasting with 7.2 million (14% of the population) saying they had downloaded a podcast and 4.4 million downloading podcasts every week – up from 1.8 million in the first survey. This compares with Arbitron/Edison's April 2008 figure of 18% of the US population using podcasts, and Pew/Internet research showing 19%

of all Internet users in the US having downloaded a podcast – up from 12% in the 2006 Pew study (Madden 2008).

RAJAR/MIDAS is also tracking how Internet listening and podcasting is affecting linear 'live' radio. Despite the fears in 2004–05 that podcasting or Internet-based radio on demand would stop people listening to linear radio, the MIDAS survey shows that, while some podcast users *are* listening to less linear radio, far more are being encouraged by podcasts to experiment and try out new programmes. In MIDAS 2, released in July 2008, 15% of podcast users said they were listening to more radio compared to 10% who said they were listening less. By MIDAS 3, released in December 2008, 35% said they were listening to live radio programmes they had not listened to before (slightly down on the previous survey), two thirds of people were listening to much the same live radio, while up to 14% were listening less and just 9% say they were listening to more live radio.

While the trends show podcasting use is growing, one of the drawbacks flagged in both the UK RAJAR and US based audience research by Arbitron/Edison and Pew/Internet is that while it gives listeners control it also requires them to make an active choice to listen. It is what Tom Webster at Edison Media Research calls 'the curse of convenience'

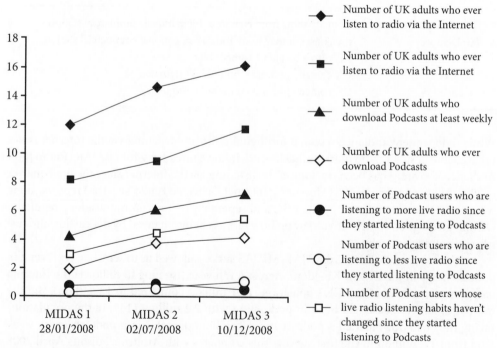

Figure 1: RAJAR/MIDAS 1–3. Podcast Trends: RAJAR/MIDAS 2008–2009.
Source: RAJAR/MIDAS 2008

(Webster 2008). On average, over 60% of podcast users say they like podcasts because they are convenient but the trends in usage show that it is also easy to postpone forever that podcast you downloaded. Of the 7.2 million podcast users in the United Kingdom, 2.3 million were lapsed users, and while 60% said they listen to a selected podcast in full, only 27% say they listen to all their subscribed podcasts. In the United States, while the Pew/Internet survey shows 19% penetration, only 17% of those using podcasts download one every day (Madden 2008). Podcasting remains a relatively new and recent activity for people, and for many it is not yet an established habit.

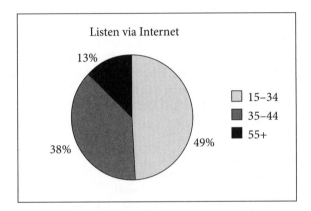

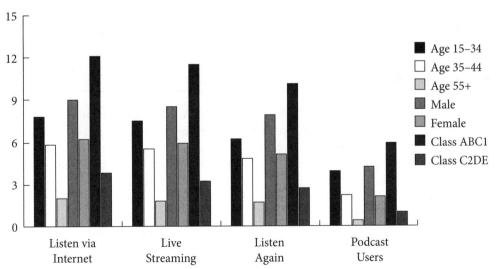

Figure 2: MIDAS demographic profiles.
Source: RAJAR/MIDAS 2008

In, effect what the research is suggesting is that while podcasting offers listeners welcome on-demand choices, the attraction and power of linear radio is that it is just there, in the background, in the kitchen, in the car, at work and that often passive consumption is likely to remain the dominant way in which listeners come across radio in the short to medium term.

What podcasting is doing is helping radio develop a life with younger audiences (under 30 years old), as demonstrated in the Danish audience studies and, according to both RAJAR/ MIDAS and the BBC's research, it is also encouraging people to experiment and try new radio shows. The profile of online and podcast users across research in the UK and US 2006–09 shows a clear bias towards young males (15–34 from higher socio-economic classes). While radio broadcasters like NPR and the BBC record high audiences for programmes with a predominately middle-aged audience, the majority of online content and podcast users are under 34 and likely to be well educated and digital media literate. While this remains the case, the emerging trend is showing increased older audiences who have more leisure time to enjoy and use online content, social networking and podcasts (Nielsen 2009).

The BBC has also commissioned research on listening online, including podcasting, (PALVIS 2008), which echoes the trends of RAJAR/MIDAS and Pew/Internet, and provides greater insight into the habits of podcast audiences. Of those downloading:

- 43% listen to podcasts once or twice a week
- Most spend no more than an hour a week listening
- Only one in ten listen for more than two hours a week
- 80% listen at home, 24% in the car and 22% on public transports
- Podcast listening peaks after 4pm
- The majority, over three quarters, listens to podcasts more than a week old

The iPod Generation grows up

In 2004, Ofcom, the UK communications regulator, published a research report on the audio habits of younger listeners under 25 called 'The iPod Generation: Devices and Desires of the Next Generation of Radio Listeners'. The term podcasting had not yet taken off, but what the report identified and described was the consumer profile which would ensure the success of podcasting as a tool. 'I want more from it … it's just a radio', complained one young male. Controlling content, personalising content, being able to pause, download and share music/ radio were all seen as important and desirable. Matching radio's future with their mobile phones was important and by 2007, 70% of all 15–24 year olds in the United Kingdom had a radio receiver on their mobile phone.[12] For them, media consumption had already come to mean via Internet, and what drove their interest was getting everything they had on the Internet in a mobile device like their phone or MP3 player. It is this generation, the digital natives as they are sometimes called (Prensky 2001), which is now growing up, who perceive and use media quite differently to their parents, and who are determining the global media landscape (RAB 2007).

If the push behind podcasting's growth has been content on demand, it has also been intrinsically linked to the penetration of high-speed broadband, Web 2.0 applications and the rapid take-off of social networking as one of the most popular activities on the Internet. It is important we see all the online trends together, as none exist in isolation; in fact, there is a clear symbiotic relationship at work. Social networking has now overtaken e-mail as the fourth most popular online activity after search and general interest portals (Nielsen 2009). Two-thirds of the world's Internet users have used social networking and it accounts for 10% – and growing – of online activity. When Apple launched the 3G iPhone, Facebook quickly became its most popular mobile application and this mobile connected user base is growing rapidly as a market force particularly in the UK and US. Facebook now has 300 million users (October 2009).

The EBU study 'Public Youth Radio in Europe' (Shulzychi 2008) captured what 15–24 year olds were doing and how it affected radio. Key findings included: young people were spending 106 minutes online a day and more than 75% visited interactive websites every month, largely to consumer audio, music and video. While young people were shifting online, their traditional radio listening had dropped by over 18% across ten years.[13] Where young people are listening to radio is via their mobile phones. By May 2009, the UK's RAJAR figures show a third of young people (24 years and under) listening to radio on their mobile phone, with mobile phone radio showing an overall 13% increase.[14]

While the early adopters and frontier digital consumers in all of this are the under 35s, Nielsen's (2009) Social Networking study shows older age groups are increasingly becoming digital converts. The biggest growth in social networking use in that study was among the 35–49 year olds, and roughly one third of Facebook users are in that age group, with a quarter over 50. The growth of this connected and inter-connected web, and the marked increased in time spent online, has helped the take-up of online and mobile content like podcasts. In a sense, podcasts have fed the growing appetite for online and mobile content, and allowed broadcast radio to swim within that converged sea.

Podcasting economics: desperately seeking revenue

'Forget squeezing million from a few megabits at the top of the charts. The future of entertainment is in the millions of niche markets at the shallow end of the bitstream.'
Chris Anderson, 'The Long Tail', *Wired*, 2004.

Podcasting has not caused radio economics to collapse. Besides the credit crunch and global economic downturn from September 2008, there are several key factors at play in the radio and media marketplace. On the one hand, advertisement revenue is bleeding from traditional media – newspapers, radio and television – to the Internet; on the other hand, the digital age has created a glut of free online content, making it increasingly difficult, though not impossible, to persuade anyone to pay for text, audio or video information or entertainment. In the United Kingdom, for example, the decline in radio commercial revenue was happening before the credit crunch and global banking crisis of 2008, but it has now left the commercial radio sector there in serious financial crisis with one analysis predicting its death except for a few podcast 'hobbyists'.[15]

Part of the problem in the UK, which is the strongest DAB consumer base in Europe with just under 30% of people owning a DAB radio, is that after a decade or more of investment in DAB stations, they have not become strong, business models. Even stations with healthy audience profiles have failed to convert those audiences into profitable revenue streams. In a sense, not unlike the newspaper sector in its shift online, the radio industry assumed digital is about changing transmission and distribution, and not about having to re-think the entire business model. The revenue model for DAB radio stations is still hanging on commercial spot advertisement rather than any innovative look at what the technology offers in terms of commerce. The move to DAB+ will help, but for some it may be too late, with more and more DAB stations switching off or cutting back. The decision by Channel 4 Radio to pull the plug on its digital radio network in October 2008 was, for some, the final sign of health, or lack of it, in the UK's digital radio market.[16]

While most podcasts are currently released for free and without advertisements, it is likely that as their usage grows and becomes more stable, they will be married with targeted advertisement and more sponsorship. One of the key aspects highlighted by the RAJAR/ MIDAS and the BBC's PALVIS reports is the public's attitude to either paying for podcasts or getting them free but with adverts. Not surprisingly 68% of users did not want to pay, but the majority 57% were prepared to accept advertisements in free content (BBC/PALVIS 2008). In RAJAR/MIDAS, only 3% had paid for content, and while micropayments may work for specific content like high value archives or exclusive content, there seems little room for micropayment in Europe since the national public broadcaster is usually prohibited under its remit from charging for content if it is in receipt of public funding. In the United Kingdom, given the extensive range of free BBC podcasts, the potential for commercial broadcasters to charge for content seems limited. NPR's ability to gain micro-payments in the United States is probably assisted by its lack of state funding and its dependence on public donations.

So while micro-payments are not a key future revenue stream for European podcasts, the likely revenue streams will be online advertisement and sponsorship embedded in the Internet websites and within the podcasts themselves. The real attraction for sponsors and advertisers with podcasts is the ability to get closer to a defined niche and create interaction with users by linking to online sales, for example. The commercial life of podcasts is closer to marketing, and much of the commercial activity around new media requires more creative, interactive marketing and sales skills than previously associated with 30-second adverts spots. The challenge is, as Tom Webster of Edison Media Research suggests, that the podcast consumer is often resistant to hard sells and direct advertisement, and far less reliable than the traditional linear radio audience (Webster 2008).

Interestingly, as the fortunes of the global newspaper industry decline, even for those who have been leaders online like *The New York Times* and *The Guardian*, the gap between the cost of content production and revenue widens, and the issue of paid content may have to be re-visited by global aggregators like Google, MySpace and YouTube, who have built their business models on providing other people's content free (Robinson 2009). Media mogul, Rupert Murdoch, whose empire includes both newspapers and new media aggregators like MySpace, has indicated that the days of free content are gone, and news will need to be paid for online as well as offline in order to correct what he calls 'a flawed business model'.[17]

That same economic trend (whereby falling revenues collapse production models and creates quality content scarcity) may also be decisive in the future of radio, particularly in relation to documentary, drama and high-end speech and music content, and may assist the development of a paid-content, online radio business model. Equally, the other lesson from the newspaper industry is that of the *Seattle Post-Intelligencer*, the 146 year old print paper which gave up the printing-press and moved completely online in spring 2009.

Weak and struggling traditional media businesses like the *Christian Science Monitor* may end up finding the only way to reduce costs is to confine distribution to the Internet and target growing online revenues and subscription content (archives, exclusive material etc) in order to survive. Music's journey from wax to vinyl to CD and MP3 has not changed music's core, nor has it altered the symbiosis of live performance and recorded music; what it has changed is music's economic model (Leonhard 2005). Chris Anderson's long tail of online content, where the sea of free content is supported by an apex of unique paid content, remains a compelling economic model for digital media, and it is as much a match for radio and podcasting as music.

Darwinian media: adapt or die

'If we are not able to make the move of content to the platforms and sites where young people increasingly spend part of their time, we risk losing both reach and share and the media users of the future.'

Alex Shulzycki, Head of SIS EBU, presentation at the
EBU Digital Radio Conference, Cagliari, September 2008.

Radio, like the rest of the media landscape, is becoming multimedia as a direct result of the digital age and convergence. Digital has not just meant migrating to new distribution methods; it also means changing production, reception and, ultimately, the way we think about and use media and communications. For radio, a portable, relatively cheap medium, the digital challenge has been to find ways of improving, enhancing and augmenting the radio experience rather than making it more expensive or more technologically complex. While much of the European industry debate on digital radio since the mid-1990s has been centred on DAB, digital terrestrial and satellite platforms, in tandem with this top down, policy-led vision has been the emerging portal of the Internet and online distribution which has generated a bottom-up revolution in how we use and create content.

While most broadcasters have had websites since the mid-1990s and launched live streaming by the late 1990s, the tipping point for radio and the Internet has been the penetration of high speed broadband and portable MP3 devices. Audio bloggers like Curry, adapting Dave Winer's Open-Source RSS technology, started a ball rolling and tapped into the growing public interest in both making their own shows and in hearing other peoples'. The blogosphere of ideas and opinions which had emerged by 2003 challenged big business and big media to pay attention and change production methods, offerings and strategies. 'Citizen Media' became a 'must have' add-on, with global media players like CNN and the BBC attempting to capture the energy behind grassroots, amateur media rather than solely compete with it. Broadcasters began to move into the new media space of social networks like YouTube and MySpace, and in the best philosophy of 'if you can't beat them, join them', began re-packaging their output online. In the end, traditional broadcasters who felt threatened by the explosion of new media content from 2003 onwards began to realise the power of their brands to build a new multimedia presence, both on and offline.

In a sense, broadcasters began the digital journey thinking they could define it. But digital is about change – disruptive change – and the media itself has been re-shaped, not by affixing its vision to technology, but by how people have adopted and used technology to be informed, entertained and communicate with one another. That process is still fluid, still clay in motion, and the re-configured media landscape is emerging out of it.

For radio, the open portal of the Internet has encouraged a re-invention of form, allowing it to hold the best of its analogue identity; portable, cheap and accessible, but adding new facets which let us save and share what we love, when and where we want to. The essence of radio remains the same but the advent of podcasting augments the experience. While all radio distribution, both linear and non linear, may eventually go via broadband, the most likely digital future is multi-platform and multimedia, with radio users enjoying the best of radio through whatever means it finds of reaching them (Marks 2006). Increasing the base of all content and communications will be social networking, with radio or online radio models emerging which marry the 'like it, share it' principle of content with traditional radio, allowing listeners to both generate their own stations or playlists as well as 'cut and paste' linear FM radios in a web based format.[18]

Podcasting as a new media radio experience can not replace linear radio, but combined with live streaming on the Internet and personalised radio tools like Last Fm and Pandora, it will play a significant part in shaping radio's digital future. If one third of the UK, to take one country as a example, is currently listening to some form of radio online and 14% of them use podcasts, it is not difficult to project that in a few years time, particularly as the under-34 year olds who have grown up on the Internet become the decision-makers, the online platform will be a significant part of how radio is received.

The question will be how radio manages to pay for itself in this interim period as audiences shift online, and as traditional media revenue dissipates into a wider pool of non-content creators like Google, YouTube and Facebook. Radio, as the examples of NPR and the BBC show, is absorbing the disruptive wave of change that online brings and adapting to it, potentially making it more fit to survive in a multimedia age. In the end, its strongest card will be holding the attention and affection of users, delivering relevant content that is easy to access, and which makes audiences listen and come back for more.

Notes

1. Forgan, Liz. BBC's Director of Radio, at the switch on of BBC DAB services, September 25 1995, promised listeners 'superb quality sound, a strong and fade free signal, a whole range of new services on simple and easy to use sets'. Retrieved on EuroDab's first newsletter, November 1995.
2. See for instance (Green, Lowry et al. 2005; Newitz 2005). For a critical discussion of the new podcasting revolution as radio, see also Berry (2006).
3. Paul Byrne, CEO Radio Kerry in interview with the author in April 2009 for Irish digital radio research project.
4. This section compiled by Per Jauert.
5. Presentation by A. Shulzychi *EBU Digital Radio Conference*, Cagliari September 2008, analysis of EBU Public Youth Radio in Europe study, June 2008.
6. DRACE survey of cross range of public broadcasters conducted by the author August-September 2008.
7. Prag, S. Executive Producer, BBC Interactive, the DRACE podcasting online survey September 2008 conducted by the author.
8. Farnon, K. RNW, DRACE podcasting online survey, August 2008 conducted by the author.
9. Aalto, T. YLE, DRACE podcasting online survey March 2009 conducted by author.
10. Russell, P. Senior Producer, RTÉ Radio Online, DRACE podcasting online survey, August 2008, conducted by the author.
11. Russell, P. update interview with author April 2009, the scale range between 6–900,000 monthly downloads relates to two different metrics. There is no independent audience audit of podcasting in Ireland.
12. The RAJAR results, May 6th 2009 showed mobile phone listening growing significantly. Radio via mobile phone grew 13% in a year and 30% of under 24s said they listened via mobile phone. ww.rajar.co.uk
13. EBU Youth Radio in Europe, (June 2008). Taking Radio France as a sample the losses fell to 30% when the age profile was extend to 15–34. This study is based on the youth services only – so reflect a decline in radio listening specifically targeted at this age group.
14. RAJAR Q1 2009, see www.rajar.co.uk.
15. Tryhorn, Chris. 'UK commercial radio dying out', *The Guardian* newspaper, March 19th, 2009, quoting analyst Claire Enders, at the Media Guardian Changing Media Summit, on the economic crisis for radio and what she saw as digital's failure to offer a transformed proposition. See www. guardian.co.uk.
16. Channel 4, in September 2008, said its digital radio plans were shelved rather than cancelled, due to the economic crisis. At the Radio Reborn conference in London, April 27th 2009 Natalie

Schwarz head of Channel 4's digital group hedged her bets. By May 6th she herself had stood down indicating that Channel 4's digital radio life was more dead than dormant.

17. See Clark, J. *The Guardian* newspaper, May 7th 2009.' News Corp will charge for newspaper websites'. Murdoch outlined he intended to introduce charges within the year. His statements came in an online conference call May 6th 2009 on www.newscorp.com at the announcement of News Corporation's 2009 quarter returns which showed a 47% side in profits.

18. Dabbl and Jelli are current models of 100% user generated radio stations which cross online and FM while Compare My Radio is a web based portal, in beta mode, which allows users to find the FM stations which best match their music tastes by searching and comparing playlists.

References

Ala-Fossi, M. and P. Jauert (2006), 'Nordic Radio in the Digital Era', in U. Carlsson (ed.), *Radio, TV & Internet in the Nordic Countries. Meeting the Challenges of New Media Technology,* Göteborg: Nordicom, pp. 65–88.

Ala-Fossi, M., S. Lax, B. O'Neill, P. Jauert and H. Shaw (2008), 'The future of radio is still digital – but which one? Expert perspectives and future scenarios for radio media in 2015', *Journal of Radio and Audio Media,* 15:1, pp. 5–25.

Arbitron/Edison (2006), 'Arbitron Radio Listening Report. The Infinite Dial. Radio's Digital Platforms', Arbiton Inc. Edison Media Research Center.

BBC (2008), *Podcasting & Listening via the Internet Survey (PALVIS),* London: MC&A.

Berry, R. (2006), 'Will the iPod Kill the Radio Star? Profiling Podcasting as Radio', *Convergence,* 12: 2, pp. 143–162.

Bull, M. (2000), *Sounding out the City: Personal Stereos and the Management of Every Day Life,* Oxford: Berg.

Businessweek (2005), 'The Future of The New York Times', *Businessweek,* 17 January 2005, http://www.businessweek.com/magazine/content/05_03/b3916001_mz001.htm. Accessed 30 May 2009.

Dorroh, J. (2008), 'The Transformation of NPR', *American Journalism Review,* October/November 2008.

The Economist (2006), 'Podcasting will change radio, not kill it', *The Economist,* 20 April 2006, http://www.economist.com/surveys/displaystory.cfm?story_id=6794210. Accessed 30 May 2009.

European Broadcasting Union (2007), *Public Radio in Europe 2007,* Geneva: European Broadcasting Union.

European Broadcasting Union (2008), *Broadcasters & the Internet,* EBU Radio SIS report, Geneva: EBU.

Evans, D. (2009), 'As Internet Radio booms, is DAB doomed?', *www.TechRadar.com.* Accessed 10 February 2009.

Glaser, M. (2005), 'Will NPR's podcasts birth a new business model for public radio?', *Annenberg Online Journalism Review,* 29 November 2005.

Green, H., T. Lowry and C. Yang (2005), 'The New Radio Revolution', *Businessweek,* 3 March 2005, http://www.businessweek.com/technology/content/mar2005/tc2005033_0336_tc024.htm. Accessed 30 May 2009.

Hammersley, B. (2004), 'Audible Revolution: Why Online Radio is Booming', *The Guardian,* 12 February 2004.

Hirst, J. (2009), 'Irish broadcasters can not afford to ignore podcasting', Speech to *Independent Broadcasters of Ireland Annual Conference,* Dublin, 3 March 2009, www.ibireland.ie. Accessed 30 May 2009.

Jardin, X. (2005), 'Podcasting Killed the Radio Star', *Wired Magazine*, 27 April 2005, www.wired.com. Accessed 30 May 2009.

Jenkins, H. (2006), *Convergence Culture*, New York: New York University Press.

Kjær, B. and M. Plon (2006), 'Podcast – en undersøgelse af praksis og potentiale blandt unge brugere / Podcast – a Study of Practice and Potential among Young Users', DR & Aarhus University.

Kozamernik, F. and M. Mullane (2005), 'An Introduction to Internet Radio', *EBU Technical Review*, October 2005.

Kozamernik, F. (2004), 'DAB: from Digital Radio to Mobile Multimedia', *EBU Technical Review* January 2004.

Levin, J. (2008), 'Podcasting audiences to double in two years', *www.eMarketer.com*. Accessed 4 February 2008.

Madden, M. and S. Jones (2008), 'Podcast Downloading', Podcasting Data Memo, 28 August 2008, Pew Internet and American Life Project, http://www.pewinternet.org/Reports/2008/Podcast-Downloading-2008.aspx. Accessed 30 May 2009.

Marks, J. (2006), 'Radio. No Single Future', Presentation to *Asia-Pacific Institute for Broadcasting Development Conference*, September 2006, Paris: UNESCO.

Marriner, C. (2006), 'Publishers and politicians want a word in your ear', *The Guardian,* 1 February 2006.

McRedmond, L. (1976), *Written on the Wind. Irish Radio 1926–1976,* Dublin: RTÉ.

McLuhan, M. (1964), *Understanding media: the extensions of man*, London: Routledge & K. Paul.

Miller, M. (2009), 'The Road Not Taken: Vivan Schiller's circuitious route to the top of NPR', *American Journalism Review*, February-March 2009.

Morrison, S. (2005), 'Hobbyists go online – but will podcasting kill the radio star?' *The Financial Times* (UK), 26 July 2005.

Newitz, A. (2005), 'Adam Curry Wants to Make You an iPod Radio Star', *Wired.com*, 13: 03.

Nielsen (2009), *Global Faces & Networked Places – A Nielsen report on social networking's new global footprint,* March 2009, http://www.nielsen-online.com/blog/2009/03/09/global-faces-and-networked-places-a-nielsen-report-on-social-networking%E2%80%99s-new-global-footprint/. Accessed 30 May 2009.

O'Mahony, C. (2009), 'Radio industry chases new revenue streams', *The Sunday Tribune*, 8 March 2009.

Ofcom (2004), *iPod Generation- Devices and Desires of the next Generation of Radio Listeners*, Research study by the Knowledge Agency, London: Ofcom.

Prensky, M. (2001), 'Digital Natives, Digital Immigrants', *On the Horizon*, 9: 5.

Radio Advertising Bureau (2007), *Radio and the Digital Native Report: How the 15–24 year olds are using radio and what this tells us about the future of the medium.* London: Radio Advertising Bureau.

Rainie, L. and M. Madden (2005), 'Podcasting catches on', Pew Internet and American Life Project, 3 April 2005, http://www.pewinternet.org/Reports/2005/Podcasting-catches-on.aspx. Accessed 30 May 2009.

RAJAR/Ipsos MORI (2008), 'Podcasting & Radio Listening via the Internet Survey', *MIDAS (Measurement of Internet Delivered Audio Services) 1,* London: RAJAR.

RAJAR/Ipsos MORI (2008), 'Podcasting & Radio Listening via the Internet Survey', *MIDAS (Measurement of Internet Delivered Audio Services) 2,* July 2008, London: RAJAR.

RAJAR/Ipsos MORI (2008), 'Podcasting & Radio Listening via the Internet Survey', *MIDAS (Measurement of Internet Delivered Audio Services) 3,* December 2008, London: RAJAR.

Robinson, J. (2009), 'Presses grind to a halt as print passes its sell by date', *Business & Media, The Observer*, 22 March 2009.

Shaw, H. (2008), 'Podcasting and Radio', Presentation, *EBU Digital Radio Conference*, Cagliari, September 2008, www.athenamedia.ie/reports. Accessed 30 May 2009.

Shaw, H. (2006), 'The Digital Future of Radio: Radio 2016, broadcasters, economics and content in a digital age', Paper presented at *European Communications Research and Education Association (ECREA) Conference*, Amsterdam, November 2005 and *VRT Digital Radio Conference*, Antwerp, May 2006, www.athenamedia.ie/reports. Accessed 30 May 2009.

Shaw, H. (2007), 'What the audience wants', Presentation, *EBU Digital Radio Conference*, Geneva, June 2007, www.athenamedia.ie/reports. Accessed 30 May 2009.

Shulzychi, A (2008), 'Youth Radio in Europe; An Overview of EBU Members Youth Channels', Presentation to the *EBU Digital Radio Conference*, Cagliari, September 2008.

Swedish Radio & Television Authority (2008), *The Future of Radio Report*, (ref 578/2008) Stockholm, June 2008.

Thomsen, M. and L. Starklint (2006), 'Podcasting – en undersøgelse af et nyt medies betydning i hverdagen / Podcasting – a Study of the Impact of a New Medium in Everyday Life', Copenhagen: IT-University.

Webster, T. (2008), 'The Podcast Consumer Revealed 2008', *Arbitron/Edison Internet & Multimedia Study*, Edison Media Research Center, April 2008.

Webster, T. (2008), 'Podcasting: The Curse of Convenience', Edison Media Research Center, www.edisonresearch.com. Accessed 20 August 2008.

Williams, R. (1995), 'BBC switches on CD-quality radio', *The Independent*, London, 28 September 1995.

Notes on Contributors

Marko Ala-Fossi (D.Soc.Sc., University of Tampere) is Senior Lecturer in Radio at the University of Tampere, Finland. His research interests include all forms of radio and audio media, political economy and media technology. Ala-Fossi is the author and co-author of books and articles on radio and broadcasting, notably *Saleable Compromises: Quality Cultures in Finnish and US Commercial Radio*. He has also produced extensive commissioned research on radio for the Finnish Ministry of Transport and Communications and for the Finnish Communications Regulatory Authority.

Lawrie Hallett is a radio executive at Ofcom with responsibilities for broadcast radio, primarily the non-profit community radio sector. Prior to that he was a radio consultant at the Community Media Association working on its Access Radio pilot project and design and implementation of Sheffield Live! Broadcast radio facilities. He has an MA in International Relations & Development Studies from the University of East Anglia, and has conducted postgraduate research in the University of Oslo's Media Department on community radio broadcasting in Norway and the United Kingdom.

Michael Huntsberger (Ph.D., University of Oregon) is an Assistant Professor of Mass Communication at Linfield College. He is the author of *The emergence of community radio in the United States: A historical examination of the National Federation of Community Broadcasters*, and articles in *Southern Review, Journalism and Mass Communication Educator, Journal of Mass Media Ethics*, and the *Public Radio News Directors Guide*. His research focuses on mass media policy, ethics and technology. Huntsberger has worked as a producer, engineer, manager and consultant in commercial, public and community radio since 1980.

Per Jauert (Cand. phil. 1975), is an Associate Professor in Media Studies at the Department of Information and Media Studies, University of Aarhus, Denmark. Jauert's research interests include media sociology, public service media, radio history, community media, and the new digital platforms for radio production, distribution and reception. He is the author of the three chapters about radio in Jensen, K.B. (2004) *Dansk Mediehistorie* [Danish Media History] I-IV, Copenhagen: Samfundslitteratur, and he has published several articles in international scholarly journals (*Media Culture and Society, Javnost – The Public Journal of Radio Studies, European Journal of Communication Research, Journal of Radio and Audio Media*). He is currently Advisory International Editor of *Journal of Radio and Audio Media* and member of the Editorial Board of *The Radio Journal,* a member of the research network "RIPE" about public service media (Ferrell Lowe, G. and P. Jauert. (eds.) (2005) *Cultural Dilemmas in Public Service Broadcasting.* Göteborg:Nordicom), Head of the Community Communication Section of IAMCR, the International Association of Media and Communication Research.

Stephen Lax is Senior Lecturer in Communications Technology at the Institute of Communications Studies, University of Leeds, United Kingdom. Stephen Lax is a member of the Media Industries Research Centre and the Centre for Digital Citizenship. His research interests are in the social role of new communications technologies and the relationship between technology and policy. He is on the steering group of the UK Radio Studies Network. He is author of *Media and Communications Technologies: a Critical Introduction* (2009) and *Beyond the Horizon: Communications Technologies Past Present and Future* (1997), and editor of *Access Denied in the Information Age* (2001).

Lars Nyre is associate professor at the Department of Information Science and Media Studies, University of Bergen, Norway, and also holds a 20% research position at the School of Journalism, Volda University College. Nyre combines research on journalism and technology. He merges classical normative ambitions from media studies with experimental design methods from information science. Nyre is the author of several books oriented to sound in the media, notably *Sound Media. From Live Journalism to Musical Recording* (2008). Nyre is the editor of *Norsk medietidsskrift* [Norwegian Journal of Media Research]. He is chair of the research group Digital Radio Cultures in Europe (www.drace.org), originally established as part of the EU-funded COST A20 network, "The Impact of the Internet on the mass media in Europe". Nyre was an executive board member of ECREA (previously called ECA) from 2003–2008, and directed the construction of www.ecrea.eu.

Brian O'Neill (PhD, University of Dublin) is Head of the School of Media at Dublin Institute of Technology, Ireland. He has published in journals such as *Media Culture and Society, New Media and Society, Journal of Radio and Audio Media,* and the *Canadian Journal of Communication* on topics including digital broadcasting policy, arts audiences, broadcasting history and information society issues for children. He has written a number of commissioned research reports on media education (2007 for UNICEF, and the Radharc Trust) and media literacy in the public sphere (2008 for the Broadcasting Commission of Ireland). In 2009 he co-authored with Helen Shaw the report *Digital Radio for Ireland: Competing Option, Public Expectations* for the Broadcasting Commission of Ireland. He is a member of the *EU Kids Online* network (Safer Internet programme) and national contact for the project, researching the risks and opportunities young people's online behaviour can produce. He is also Interim Chair of the IAMCR Audiences Section and a member of the international editorial board of the *Journal of Radio and Audio Media* (BEA).

Paddy Scannell (Ph.D., University of Westminster) is Professor in Media, Culture and Society at the University of Michigan, Ann Arbor. Paddy Scannell worked for many years at the University of Westminster (London) where he and his colleagues established, in 1975, the first undergraduate degree program in Media Studies in the UK. He is a founding editor of *Media, Culture and Society*, which began publication in 1979 and is now issued six times yearly. He is the author of *A Social History of British Broadcasting, 1922–1939*, which he wrote with David Cardiff, editor of *Broadcast Talk* and author of *Radio, Television and Modern Life*. He is currently working on a trilogy. The first volume, *Media and Communication*, was published in June 2007. Professor Scannell is now working on the second volume, *Television and the Meaning of 'Live'*. The third volume, *Love and Communication*, is in preparation. His research interests include broadcasting history and historiography, the analysis of talk, the phenomenology of communication and culture and communication in Africa.

Helen Shaw is Managing Director of Athena Media, a digital media production company based in The Digital Hub, Dublin, Ireland. Helen Shaw is a former Managing Director of RTÉ Radio and previously was a senior editor with the BBC. Helen is a graduate of UCD (BA, MA) and of Dublin City University (Post Grad Journalism) and was an invited Fellow to Harvard University for the academic year 2002–2003 where she worked on advanced research on media globalisation and the Harvard paper 'The Age of McMedia' (Harvard, Weatherhead Center, 2003).

She has worked as a media consultant with public broadcasters SABC, RTÉ, VRT and has presented at both academic and industry conference including AIB (Association of International Broadcasting) Prague 2004, BCI, Broadcasting Commission of Ireland 2004, ECREA Amsterdam 2005, EBU Digital Radio Geneva 2007, EBU News+Current Affairs Stuttgart 2007 and EBU Digital Radio Cagliari 2008.

She has worked on media policy for the Irish Government's Department of Communications (Ox report 2004), and worked as a media consultant on the creation of a media communications agency for the development agencies in Ireland, which later became Connect World. In 2009 she completed a report on digital radio in Ireland for the Broadcasting Commission of Ireland with Brian O'Neill. She is the author of the Irish Media Directory and Guide (Gill and MacMillan 2005).

Alan G. Stavitsky (Ph.D., The Ohio State University) is Senior Associate Dean of the University of Oregon School of Journalism and Communication, and Director of the School's George S. Turnbull Portland Center. He is the author and co-author of books and articles on public broadcasting policy, history and ethics, notably *Independence and Integrity: A Guidebook for Public Radio Journalism* and *A History of Public Broadcasting*. A former broadcast journalist, Stavitsky is a consultant to public media organizations in the United States.

Index